Novel Nanoparticles and Their Enhanced Polymer Composites

Novel Nanoparticles and Their Enhanced Polymer Composites

Editors

Yuwei Chen
Yumin Xia

Basel • Beijing • Wuhan • Barcelona • Belgrade • Novi Sad • Cluj • Manchester

Editors

Yuwei Chen
School of Polymer Science
and Engineering
Qingdao University of
Science and Technology
Qingdao
China

Yumin Xia
College of Material Science
and Engineering
Donghua University
Shanghai
China

Editorial Office
MDPI
St. Alban-Anlage 66
4052 Basel, Switzerland

This is a reprint of articles from the Special Issue published online in the open access journal *Polymers* (ISSN 2073-4360) (available at: www.mdpi.com/journal/polymers/special_issues/H15I1F65KL).

For citation purposes, cite each article independently as indicated on the article page online and as indicated below:

Lastname, A.A.; Lastname, B.B. Article Title. *Journal Name* **Year**, *Volume Number*, Page Range.

ISBN 978-3-7258-0550-1 (Hbk)
ISBN 978-3-7258-0549-5 (PDF)
doi.org/10.3390/books978-3-7258-0549-5

© 2024 by the authors. Articles in this book are Open Access and distributed under the Creative Commons Attribution (CC BY) license. The book as a whole is distributed by MDPI under the terms and conditions of the Creative Commons Attribution-NonCommercial-NoDerivs (CC BY-NC-ND) license.

Contents

About the Editors vii

Preface ix

Quan Wang, Junbo Che, Weifei Wu, Zhendong Hu, Xueqing Liu and Tianli Ren et al.
Contributing Factors of Dielectric Properties for Polymer Matrix Composites
Reprinted from: *Polymers* **2023**, *15*, 590, doi:10.3390/polym15030590 1

Tong Zheng, Wenjing Jia, Hongjie Meng, Jiajie Li and Xundao Liu
Synthesis of Monodisperse Mesoporous Carbon Spheres/EPDM Rubber Composites and Their Enhancement Mechanical Properties
Reprinted from: *Polymers* **2024**, *16*, 355, doi:10.3390/polym16030355 22

Guangyong Liu, Huiyu Wang, Tianli Ren, Yuwei Chen and Susu Liu
Systematic Investigation of the Degradation Properties of Nitrile-Butadiene Rubber/Polyamide Elastomer/Single-Walled Carbon Nanotube Composites in Thermo-Oxidative and Hot Oil Environments
Reprinted from: *Polymers* **2024**, *16*, 226, doi:10.3390/polym16020226 35

Xiaojun Ma, Dongxu Mao, Wenkai Xin, Shangyun Yang, Hao Zhang and Yanzhu Zhang et al.
Flexible Composite Electrolyte Membranes with Fast Ion Transport Channels for Solid-State Lithium Batteries
Reprinted from: *Polymers* **2024**, *16*, 565, doi:10.3390/polym16050565 50

Haibin Li, Shisheng Zhou, Shanxiang Han, Rubai Luo, Jingbo Hu and Bin Du et al.
Thermoelectric Properties of One-Pot Hydrothermally Synthesized Solution-Processable PEDOT:PSS/MWCNT Composite Materials
Reprinted from: *Polymers* **2023**, *15*, 3781, doi:10.3390/polym15183781 62

Zhibin Wang, Zhanfeng Hou, Xianzhen Liu, Zhaolei Gu, Hui Li and Qi Chen
Preparation of Zinc Oxide with Core–Shell Structure and Its Application in Rubber Products
Reprinted from: *Polymers* **2023**, *15*, 2353, doi:10.3390/polym15102353 72

Jun He, Baoyuan Huang, Liang Wang, Zunling Cai, Jing Zhang and Jie Feng
Enhancing Natural Rubber Tearing Strength by Mixing Ultra-High Molecular Weight Polyethylene Short Fibers
Reprinted from: *Polymers* **2023**, *15*, 1768, doi:10.3390/polym15071768 83

Zhanfeng Hou, Dawei Zhou, Qi Chen and Zhenxiang Xin
Effect of Different Silane Coupling Agents In-Situ Modified Sepiolite on the Structure and Properties of Natural Rubber Composites Prepared by Latex Compounding Method
Reprinted from: *Polymers* **2023**, *15*, 1620, doi:10.3390/polym15071620 94

Lijian Xia, Anmin Tao, Jinyun Cui, Abin Sun, Ze Kan and Shaofeng Liu
ESBR Nanocomposites Filled with Monodisperse Silica Modified with Si747: The Effects of Amount and pH on Performance
Reprinted from: *Polymers* **2023**, *15*, 981, doi:10.3390/polym15040981 113

Peipei Sun, Ziwen Zhou, Licong Jiang, Shuai Zhao and Lin Li
The Study of Enteromorpha-Based Reinforcing-Type Flame Retardant on Flame Retardancy and Smoke Suppression of EPDM
Reprinted from: *Polymers* **2022**, *15*, 55, doi:10.3390/polym15010055 132

About the Editors

Yuwei Chen

Prof. Dr. Yuwei Chen works at the School of Polymer Science and Engineering, Qingdao University of Science and Technology. His research interests include polymer composites, dielectric materials, UV-curable resins, fibers, surface treatment, polymer chemistry and material manufacturing.

Yumin Xia

Prof. Yumin Xia works at Donghua University. His reserch interests include high-performance fibers, surface treatment technology, functional polymer composites, 3D printing and funcional particles.

Preface

The addition of functional fillers into a polymer matrix has been extensively explored and used in many applications, among which are electronics, medicine, aerospace, energy storage, sensors, etc. With the fast development of science and technology, novel particles and their composites with multiple functions have been invented to meet new requirements. Moreover, new manufacturing methods for preparing the particles and composites are also emerging. One challenge is to prepare high-performance or functional polymer composites with a low filler content employing an easy, scale-up approach. Another challenge is to integrate the function into the composite efficiently and subtly through a rational design of the particles or the particle distribution in the polymer matrix. This reprint aims to highlight the advances and cutting-edge technologies of particles and particle-reinforced functional polymer composites. Original research articles and reviews are included and the research areas encompass the synthesis and characterization of novel particles, novel manufacturing technology for particle-reinforced polymer composites, rational design of the distribution of particles in a polymer matrix, properties of composites enhanced by particles and multifunctional polymer composites.

Yuwei Chen and Yumin Xia
Editors

Review

Contributing Factors of Dielectric Properties for Polymer Matrix Composites

Quan Wang [1], Junbo Che [1], Weifei Wu [1], Zhendong Hu [1], Xueqing Liu [2], Tianli Ren [3], Yuwei Chen [1,*] and Jianming Zhang [1]

[1] Key Laboratory of Rubber-Plastics, Ministry of Education/Shandong Provincial Key Laboratory of Rubber-Plastics, Qingdao University of Science & Technology, Qingdao 266042, China
[2] Key Laboratory of Optoelectronic Chemical Materials and Devices, Ministry of Education and Flexible Display Materials and Technology Co-Innovation Centre of Hubei Province, Jianghan University, Wuhan 430056, China
[3] Mississippi Polymer Institute, The University of Southern Mississippi, Hattiesburg, MS 39406, USA
* Correspondence: yuweichen@qust.edu.cn

Abstract: Due to the trend of multi-function, integration, and miniaturization of electronics, traditional dielectric materials are difficult to satisfy new requirements, such as balanced dielectric properties and good designability. Therefore, high dielectric polymer composites have attracted wide attention due to their outstanding processability, good designability, and dielectric properties. A number of polymer composites are employed in capacitors and sensors. All these applications are directly affected by the composite's dielectric properties, which are highly depended on the compositions and internal structure design, including the polymer matrix, fillers, structural design, etc. In this review, the influences of matrix, fillers, and filler arrangement on dielectric properties are systematically and comprehensively summarized and the regulation strategies of dielectric loss are introduced as well. Finally, the challenges and prospects of high dielectric polymer composites are proposed.

Keywords: polymer composites; dielectric properties; microstructure; functional polymers

Citation: Wang, Q.; Che, J.; Wu, W.; Hu, Z.; Liu, X.; Ren, T.; Chen, Y.; Zhang, J. Contributing Factors of Dielectric Properties for Polymer Matrix Composites. *Polymers* **2023**, *15*, 590. https://doi.org/10.3390/polym15030590

Academic Editor: Serge Bourbigot

Received: 8 January 2023
Revised: 21 January 2023
Accepted: 22 January 2023
Published: 24 January 2023

Copyright: © 2023 by the authors. Licensee MDPI, Basel, Switzerland. This article is an open access article distributed under the terms and conditions of the Creative Commons Attribution (CC BY) license (https://creativecommons.org/licenses/by/4.0/).

1. Introduction

As the demand for electronics and capacitor devices increases, high dielectric materials have attracted increasing attention [1–3]. Polymer materials own advantages of ease of processing, flexibility, and good mechanical properties but the dielectric properties are usually less than satisfactory. Therefore, the preparation of high dielectric composites by introducing high dielectric fillers has become a research hotspot [4,5]. Fillers or interfaces can be easily polarized under external electric fields then enhances the dielectric permittivity of polymer composites [6]. Compared with traditional dielectric materials, high dielectric polymer composites offer more benefits, such as easy processing, excellent mechanical properties, and good flexibility [7].

The dielectric permittivity (ε) and dielectric loss ($\tan\delta$) composes the dielectric properties of the composites [8]. Additionally, the ε is composed by a real (ε') and imaginary part (ε''). The dielectric loss refers to the phenomenon of heat generation accompanied by energy consumption during polarization [8–10]. The relationship between the dielectric permittivity (ε) and dielectric loss ($\tan\delta$) is shown by the following Equation (1):

$$tan\delta = \frac{\varepsilon''}{\varepsilon'} \qquad (1)$$

where ε' is the real permittivity of the system and ε'' is the imaginary permittivity of the system. Dielectric loss is mainly attributed to polarization loss and conductivity loss. Polarization loss is mainly generated by the polarization of the molecular dipole. Polarization loss occurs during the polarization and relaxation process, which inevitably consumes electrical energy to overcome internal viscous resistance of the medium, thus resulting in dielectric loss of the material. Any movement of current carriers, even in very restricted areas, also consumes energy to overcome the resistance and results in conductivity loss [6]. Both the dielectric permittivity and dielectric loss directly affect the practical application of dielectric materials [11,12]. By the reasonable selection principle of the matrix and filler, dielectric properties of composites are manipulated. For instance, the addition of conductive fillers or ceramic fillers can increase the dielectric permittivity. By introducing insulating fillers or core-shell structures to block the formation of conductive paths, the increase of dielectric loss can be effectively suppressed [13–15]. Of course, not all fillers can enhance dielectric permittivity, the dielectric permittivity decreases when fillers such as POSS with a cage structure are introduced [16–18].

Based on physics theory, polarization can affect the dielectric permittivity of the material and the charge of the material accumulates under the external electric filed leading to the polarization phenomenon (Figure 1). Factors affecting dielectric permittivity include electron polarization, atomic polarization, dipole polarization, and interfacial polarization [6,19,20]. Both electronic polarization and atomic polarization are collectively referred to as deformation polarization or induced polarization. Polymers are subject to deformation polarization or induced polarization in the high-frequency region. Interfacial polarization is generated by the aggregation of electrons or ions in the dielectric at the non-homogeneous interface and due to the different polarization rates of the components on either side of the interface, this often occurs at the interfaces of impurities, defects, crystalline, and amorphous regions [21–24]. Compared to the three polarization phenomena mentioned above, dipole polarization usually takes a longer time since the molecules are required to override inertia and resistance during polarization; therefore, dipole polarization occurs at a low-frequency range. The above mentioned four types of polarization determine the dielectric permittivity of the material. The relationship between polarization and dielectric permittivity is as follows:

$$P = (\varepsilon_r - 1)\varepsilon_0 E \tag{2}$$

where P is the polarization intensity and ε_r and ε_0 represents the dielectric permittivity of the material and vacuum, respectively. E is the strength of applied electric fields. It can be derived that the polarization intensity and dielectric permittivity are proportional to each other based on Equation (2). Hence, the higher the polarity of the material the higher its dielectric permittivity. This is also the cause of why the polymer owning of a large number of polar groups are chosen as the matrix of high dielectric polymer composites. Table 1 shows dielectric properties of common matrix at 1 KHz frequency [6].

Table 1. Dielectric properties of various polymers [6].

Polymer Materials	Dielectric Permittivity (1 KHz)	Loss Tangent (1 KHz)	References
Polytetrafluoroethylene (PTFE)	2	0.0001	[25]
Biaxially oriented polypropylene (BOPP)	2.2	0.0002	[26]
Low-Density Polyethylene (LDPE)	2.3	0.003	[25,27]
High-Density Polyethylene (HDPE)	2.3	0.0002–0.0007	[25,27]
Polystyrene (PS)	2.4–2.7	0.008	[28]
Polydimethyl siloxane (PDMS)	2.6	0.01	[25,29]
Polycarbonate (PC)	3.0	0.0015	[28]
Polyvinyl chloride (PVC)	3.4	0.018	[30]
Polyimide (PI)	3.5	0.04	[31]
Polyethylene glycol terephthalate (PET)	3.6	0.01	[25,32]
Poly(ether-ether-ketone) (PEEK)	4.0	0.009 (100 KHz)	[25,33,34]
Epoxy	4.5	0.015	[25,28,35,36]
Polymethyl methacrylate (PMMA)	4.5	0.05	[25,37–39]
Polyurethane (PU)	4.6	0.02	[40]
Polyvinylidene difluoride (PVDF)	10	0.04	[36]
Polyvinyl alcohol (PVA)	12	0.3	[41,42]

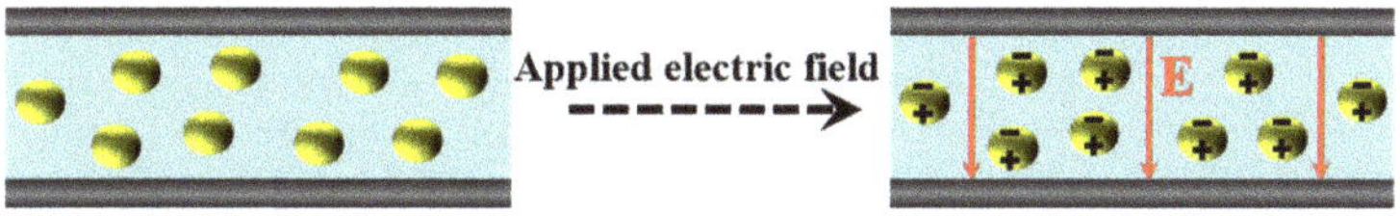

Figure 1. Polarization under electric fields.

Compared with the principle of matrix selection, the selection strategy of fillers for high dielectric polymer composites have to take more factors into consideration, including polarity, electrical conductivity and interfacial effects, processing properties, and mechanical properties [43]. The commonly used fillers are classified as the following: (1) conductive fillers [44–47]; (2) ceramic fillers; and (3) polar polymer fillers [48–55]. The addition of fillers can increase the dielectric constant of composites to some extent; however, it also increases the dielectric loss simultaneously, which is undesirable for practical applications.

In addition to the selection strategies for fillers, controlling the distribution of fillers by external fields can also improve the dielectric permittivity of the composite. External fields can align the particles by shearing force or electrophoretic force along one direction in the polymer matrix (Figure 2). This alignment structure effects the dielectric properties on many aspects. Based on the above considerations, this review systematically discusses the impact on fillers, structural design on dielectric permittivity of composites, as well as regulation strategies of dielectric loss. Typically used fillers of a different nature, such as conductive, inorganic, and organic, have been investigated. At the end, the challenges and prospects of high dielectric polymer composites are proposed [55–58]. The dielectric properties of polymer high dielectric composites are mainly affected by the matrix and filler. Distinct from reported reviews, in this review, we introduce not only the effects of polymer matrix and filler type but also the cutting-edge research, such as the effect of the distribution method of filler on the dielectric properties of composites and the up-to-date method of suppressing the dielectric loss of the composites. We Hope that this review article will give readers a comprehensive understanding and inspire future multidisciplinary research efforts in high dielectric polymer composites arena.

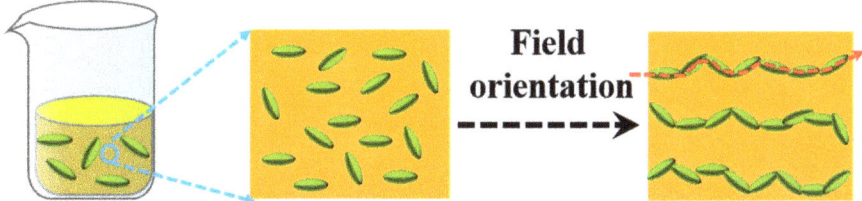

Figure 2. Fillers oriented under external electric fields.

2. Influence of Polymer Matrix on Dielectric Properties of Composites

Polymer materials are widely used in electronic fields, such as integrated circuit boards, film capacitors, and display screens, due to their excellent processability and low dielectric loss. Due to the huge demand, a variety of polymer materials have gradually emerged, including polystyrene, polyethylene, polycarbonate, polyvinylidene fluoride, and other materials [59–62].

As shown in Table 1, polymers with high polarity are preferred for high dielectric polymer composites matrix [6]. The most commonly used include PVDF, PMMA, etc., while polyvinyl alcohol has a high dielectric permittivity, but its loss is also very high, affecting the practical application, so it is generally not used as a matrix. PTFE, BOPP, LDPE, HDPE, and PS have high molecular regularity and there are no polar groups in the molecular chain. Therefore, these materials own low dielectric permittivity and dielectric loss. The polar groups in the system will improve the dielectric permittivity and dielectric loss of the composite, such as PVDF, PVA, etc. Of course, the same material's dielectric permittivity and dielectric loss are not invariable; the different test frequency will also cause the change of dielectric permittivity. Figure 3 shows that different polarization phenomena will occur at different test frequencies and accumulate continuously [6]. Therefore, dielectric permittivity and dielectric loss for a certain test frequency may be a result of multiple polarization phenomena superimposed.

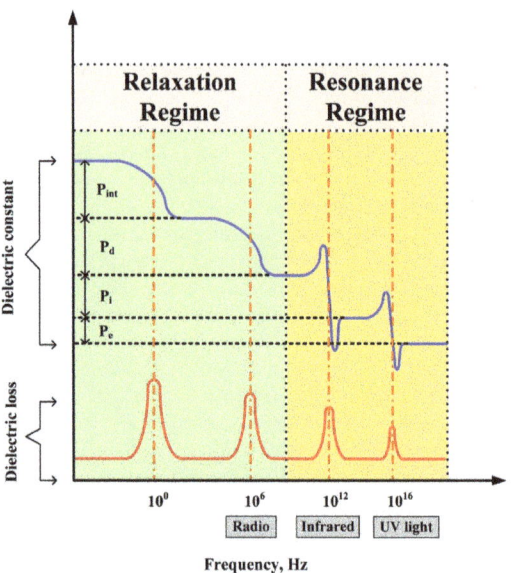

Figure 3. Polarization types at different frequencies. P_e, P_i, P_d, and P_{int} refer to e-polarization, ion-polarization, dipole-polarization, and interfacial polarization, respectively. Adapted with permission [6]. Copyright 2016, American Chemical Society.

The polymer matrix can provide excellent mechanical properties for the composite and also has a great influence on the dielectric properties of the composite. The dielectric properties of composite materials are greatly influenced by the matrix [63,64]. As shown in Figure 4, the dielectric composite with the middle layer of polymethyl methacrylate (PMMA) and the outer layer of polyvinylidene fluoride-co-hexafluoropropvlene (PVDF-HFP) filled with BaTiO$_3$ nanoparticles (BT-NPs) was prepared (Figure 4a,b) [65]. The dielectric properties of PMMA, PVDF-HFP, PVDF-HFP/PMMA/PVDF-HFP, and the designed sandwich composites (PVDF-HFP/BT-NPs)/PMMA/ (PVDF-HFP/BT-NPs) were characterized. The dielectric permittivity of P(VDF-HFP) was around 11 (at 100 Hz), while that of PMMA was only 4. The dielectric permittivity of the sandwich structure composite is 4.8 (Figure 4c). The dielectric permittivity of the sandwich structure composite could be enhanced to around 6 by adding 20 wt% BaTiO$_3$ nanoparticles, but the dielectric permittivity of PVDF-HFP was still higher than that of the composite. In addition, the addition of BaTiO$_3$ nanoparticles into PMMA also caused the increase of dielectric loss (Figure 4d), which is undesirable. Therefore, the effect of polarity on the dielectric permittivity of the polymer is pronounced. There are a large number of polar groups in the PVDF matrix, and PVDF has a higher dielectric constant than PMMA and the difference is very large. In other words, a polymer matrix with weak polarity is not always able to exceed a polymer matrix with strong polarity by adding a certain content of filler. Therefore, the selection of matrix is very important for high dielectric composites [63,64]. Commonly, polymers with strong polarity are preferred when designing high dielectric composites.

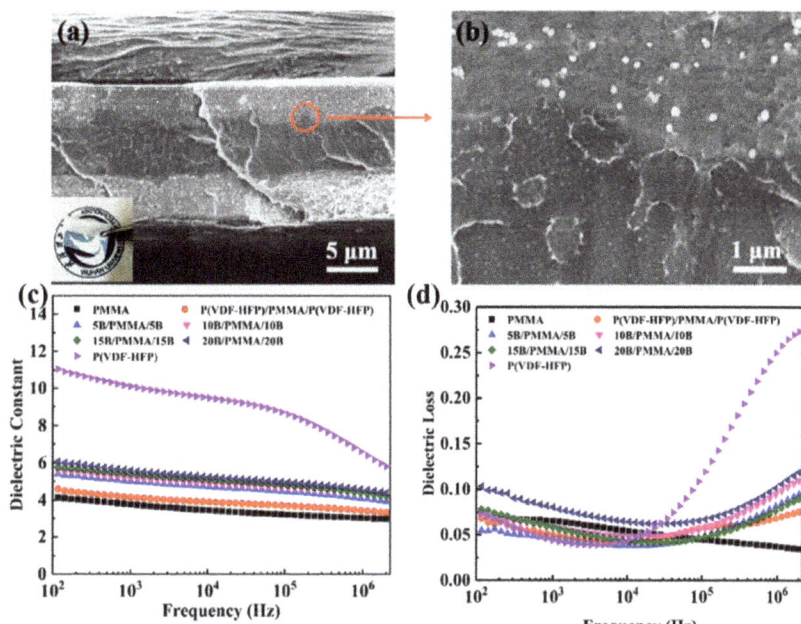

Figure 4. (**a**) SEM diagrams for sandwich structure composite with 10 wt% BT-NPs. (**b**) SEM images for adjacent region of PVDF-HFP/BT-NPs and PMMA. The dielectric permittivity (**c**) and dielectric loss (**d**). Adapted with permission [65]. Copyright 2021, Elsevier.

3. Influence of Fillers on Dielectric Properties of Polymer Composites

Filler has a great influence on the performance of composites as well. Usually, fillers with high polarity or conductivity can be added to greatly increase the dielectric permittivity of the composite. However, the dispersion is crucial to both the dielectric permittivity and dielectric loss [66–69]. Poor dispersion leads to higher dielectric loss and lower dielectric permittivity. Moreover, the addition of filler will inevitably generate a mass of interface in the composite, which leads a troublesome issue of increasing dielectric loss [70–72]. It is worth mentioning that the interaction between the filler and the matrix may restrict the movement of the molecular chains, which can reduce dielectric loss [73]. In high dielectric composite systems, conductive fillers, ceramic fillers, and polar polymer fillers are the mainly used categories. Their advantages and disadvantages were summarized in the Table 2.

Table 2. Common high dielectric fillers and their advantages and disadvantages.

Type of Filler	Common Materials	Advantages	Disadvantages	References
Conductive fillers	Ag, Au, CNTs, Graphene, etc.	High dielectric	Difficult to disperse; Poor compatibility	[44–47]
Ceramic fillers	$BaSrTiO_3$, $BaTiO_3$, $CaCuTiO_3$, etc.	High dielectric	High dielectric loss; Poor compatibility	[48–50]
Polar polymers	CNCs, PAN, etc.	Good compatibility	Limited increase of dielectric permittivity	[51–55]

3.1. Application of Conductive Fillers in Dielectric Composites

Conductive fillers are mainly divided into metallic materials and carbon materials. Metallic particles include silver, copper, aluminum, and nickel and carbon materials involve carbon nanotubes, graphene, carbon black, etc. [74–78]. Although all these particles above mentioned can greatly increase the dielectric permittivity of composites, their disadvantages cannot be ignored. On the one hand, the conductive fillers are less compatible with polymer matrix, accompanying agglomeration phenomenon, which has bad influence on both processing and performance of the composites [10]. On the other hand, the fillers contact each other and form conductive pathways easily due to their good conductivity, which results in current leakage and increases the dielectric loss of the composites [43]. Therefore, many researchers have addressed the drawbacks of poor filler-matrix compatibility by modifying conductive fillers. For instance, the compatibility problem can be solved by modifying the conductive filler with a insulative polymer layer, and if the layer is isolating, the formation of the conductive pathway can also be blocked [79,80].

Chen et al. prepared carboxylated multiwalled carbon nanotubes using acid oxidation and then acyl chlorinated carbon nanotubes (NH_2-MWNT) were prepared by immersing the carboxylated multiwalled carbon nanotubes into chlorinated sulfoxide [81]. Then, NH_2-MWNT/PI composites were prepared by using modified carbon nanotubes as fillers and polyimide (PI) as the matrix (Figure 5a,b). The dielectric permittivity of composites gradually increases with increasing NH_2-MWNT content, which could reach up to 31 (1 KHz) until the content of NH_2-MWNT was 10 wt%, and the dielectric loss is only 0.022. The dielectric permittivity of the composites began to decrease when the content of NH_2-MWNT was over 10 wt%. That is because NH_2-MWNT has good dispersion in the composite by surface modification when the content of NH_2-MWNT was not exceeding 10 wt%, while the dispersion of NH_2-MWNT became worse as long as the content was over the limit, which leads to a decrease in dielectric permittivity and an increase in dielectric loss (Figure 5c–e). Carbon nanotubes have excellent electrical conductivity, which can increase the polarization phenomenon and the dielectric constant of composites. The reduction of mutual contact of carbon nanotubes after modification can suppress the elevated dielectric loss to some extent. Therefore, the method of modifying conductive fillers by polymer materials can improve the compatibility between the conductive fillers and polymer matrix to a certain extent and also can restrain the formation of conductive pathways.

3.2. Application of Ceramic Fillers in Dielectric Composites

Ceramic materials have excellent dielectric properties, but they suffer from poor processing properties due to their low impact resistance and high brittleness, which limit their use in electronics [82]. Therefore, ceramic fillers are commonly applied to prepare high dielectric composites, which own processing and mechanical properties of the polymer matrix while possessing excellent dielectric permittivity of fillers [83–85]. However, to improve the dielectric permittivity of composites, a high loading of ceramic fillers is often necessary and the poor compatibility of ceramic fillers with polymers often introduces a large number of defects and voids when the loading is high, which will directly increase dielectric loss and greatly affect mechanical properties of composites. The ceramic filler has a great enhancement of the dielectric permittivity of the composite. Costa et al. used $BaTiO_3$ as fillers in a silk fibroin matrix to prepare bio-based composites with high dielectric properties (Figure 6) [86]. The distribution of $BaTiO_3$ particles in silk fibroin are relatively uniform. The dielectric permittivity of the composites improved from 4 to 141 at 1 KHZ for $BaTiO_3$ addition of 40 wt%. While the dielectric loss also increased from 0.1 to 10, this seriously affects the practical application since high dielectric loss means lot of heat will be generated during use. $BaTiO_3$ has very high polarizability, which can greatly improve the dielectric constant of composites. However, as a ceramic filler, the dielectric loss will also greatly increase, which will seriously affect the practical application of polymer high dielectric composites.

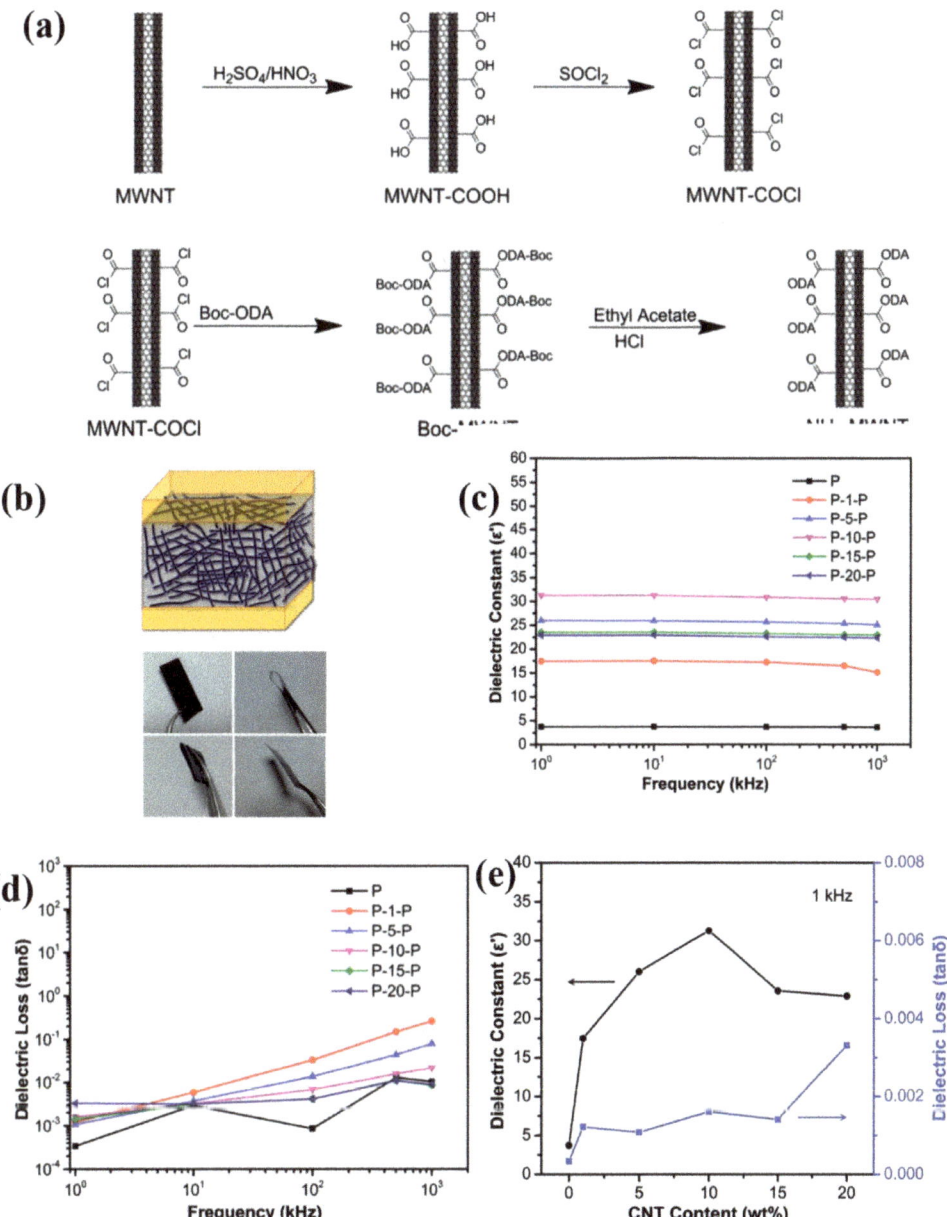

Figure 5. (**a**) Synthetic route of NH$_2$-MWNT. (**b**) Schematic and physical diagram of NH$_2$-MWNT/PI composites. (**c**) Dielectric permittivity and (**d**) dielectric loss of NH$_2$-MWNT/PI composite films. (**e**) The dielectric properties of the composites at 1 KHz influenced by the NH$_2$-MWNT content. Adapted with permission [81]. Copyright 2014, Royal Society of Chemistry.

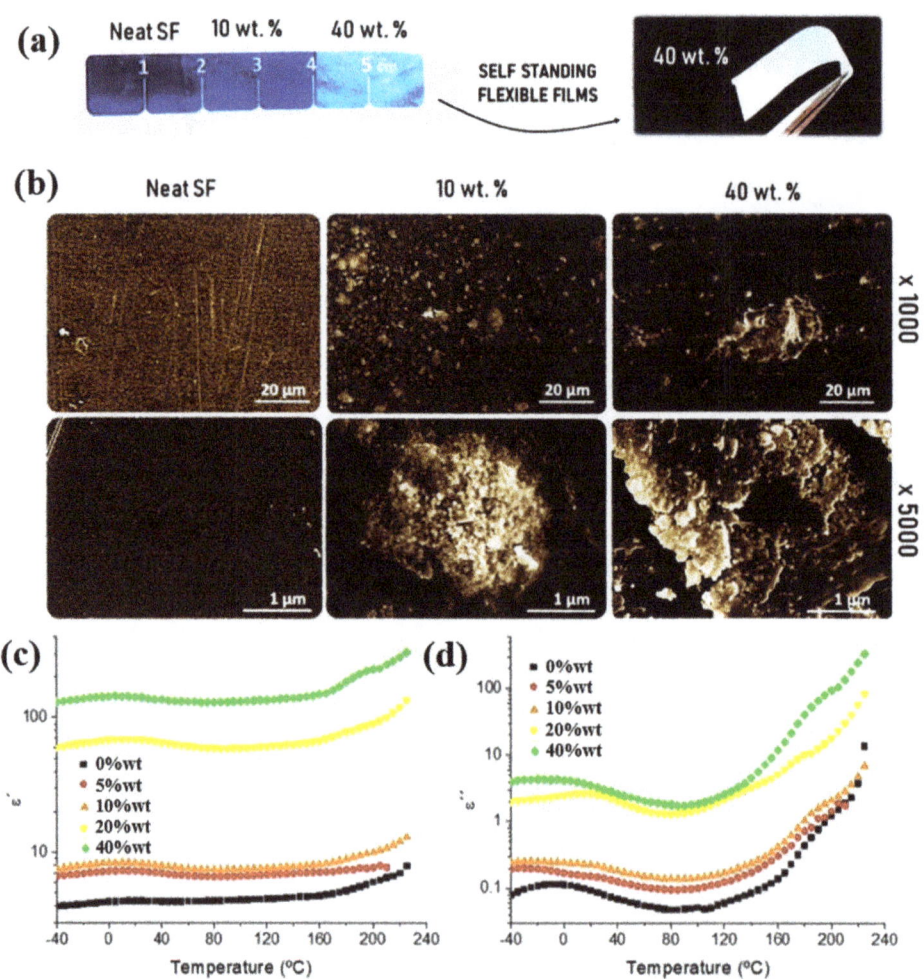

Figure 6. (a) Images of flexible silk fibroin/BaTiO$_3$ composite films. (b) Cross-sectional morphology of silk fibroin films, 10 wt% BaTiO$_3$/silk fibroin, and 40 wt% BaTiO$_3$/silk fibroin composites. Variation of ε' (c) and ε'' (d) with temperature for BaTiO$_3$/silk fibroin composites at 1 KHz. Adapted with permission [86]. Copyright 2021, Elsevier.

To address the high dielectric loss and poor compatibility of ceramic fillers, many researchers have explored strategies to increase the interaction between a ceramic filler and the matrix [87–89]. Yang et al. developed a method of preparing vinylated BaTiO$_3$ [90]. Both polystyrene modified BaTiO$_3$ (PS@BaTiO$_3$) and poly (methyl methacrylate) modified BaTiO$_3$ (PMMA@BaTiO$_3$) are investigated systematically, as shown in Figure 7. BaTiO$_3$ was modified by polymer and the dielectric loss remained very low (<0.04) when the dielectric permittivity was increased above 30 at 100 Hz. BaTiO$_3$ often has uneven dispersion phenomenon when used as filler, which is one of the reasons for the high dielectric loss of composites with BaTiO$_3$ as filler. Therefore, modifying BaTiO$_3$ with polymer materials can improve the compatibility with the matrix and achieve the purpose of suppressing the dielectric loss.

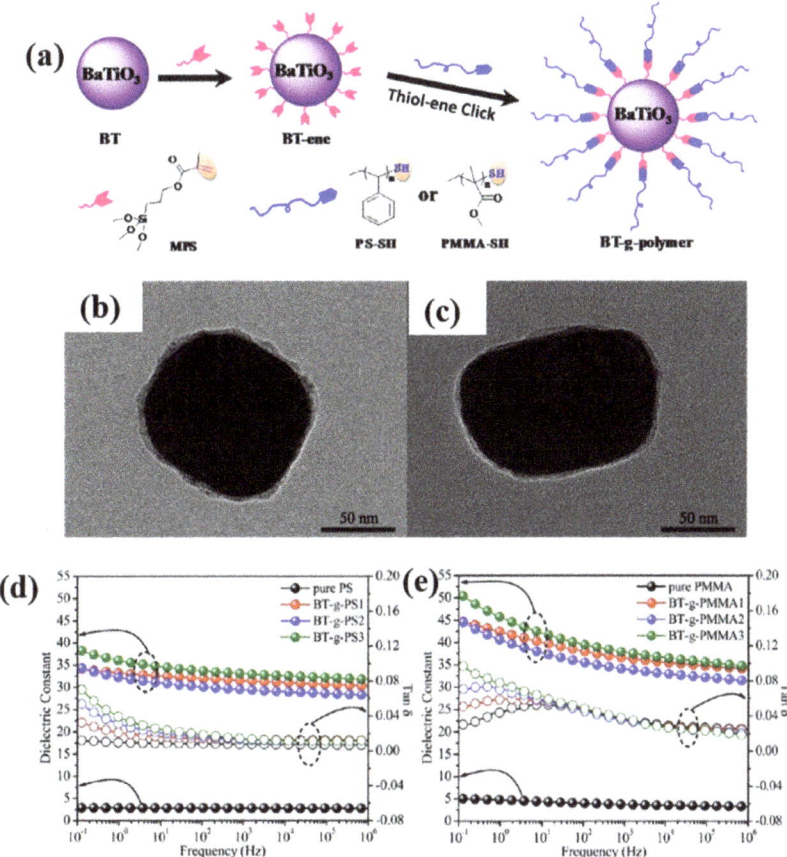

Figure 7. (a) Preparation of core-shell structured polymer@BaTiO$_3$ Nanoparticles. (b) PS@BaTiO$_3$ nanoparticle and (c) PMMA@BaTiO$_3$ at high magnification. Dielectric properties of PS@BaTiO$_3$ (d) and PMMA@BaTiO$_3$ (e) composites. Adapted with permission [90]. Copyright 2014, American Chemical Society.

3.3. Application of Polymer Fillers in Dielectric Composites

Some polar polymers cannot be used as a matrix for high dielectric polymer composites directly, but they are effective as filler to increase the dielectric properties of the composite due to the very high content of polar groups. For instance, both polyaniline (PANI) and cellulose nanocrystals (CNCs) as fillers can effectively improve the dielectric properties of composites [91–93].

Polyaniline is a special conductive polymer material. As a filler, it has good compatibility with the matrix and can also improve the dielectric permittivity of composites. Dash et al. has successfully enhanced the dielectric properties of thermoplastic polyurethane (TPU) using polyaniline (PANI) as a filler [94]. As shown in Figure 8, both ε' and ε'' of the composites increase with the addition of PANI. When the amount of PANI is less than 15 wt%, the break strength of the composite increased. PANI as a special conductive polymer can not only improve the dielectric constant of the composites, but also PANI has excellent compatibility with the polymer matrix when added to the polymer at a certain content and has a low impact on the dielectric loss of the composite (as shown in Figure 8c). Wu et al. prepared PANI/PDMS composites by orienting PANI fillers in a polydimethylsiloxane (PDMS) matrix through AC electric fields [91]. When the PANI addition was 10%, the dielectric permittivity of PANI/PDMS (random) increased by 1.72 at 100 Hz and that of PANI/PDMS (aligned) increased by 96.02 at 100 Hz. In addition to being used as a filler, many researchers use PANI to modify the filler to prepare composite materials with excellent dielectric properties. Zhang et al. prepared composites with excellent dielectric properties by embedding polyaniline modified $BaTiO_3$ (BT@PANI) as fillers into polyvinylidene cohexafluoropropylene [P(VDF-HFP)] [95]. With 20 vol% of BT@PANI, the dielectric permittivity of the composite can reach 99.1 at 1 KHz, which is 83 higher than that of 20 vol% BT (16.1) and 88.8 higher than that of the original P(VDF-HDP) (10.3). In the study of Rahnamol A. M. et al., PANI and GO hybrid materials were used as fillers to increase the dielectric properties of an epoxy resin matrix [96]. In the work of many researchers, PANI not only provides excellent dielectric properties for composite materials but also has better compatibility with the matrix and thus results in better mechanical properties.

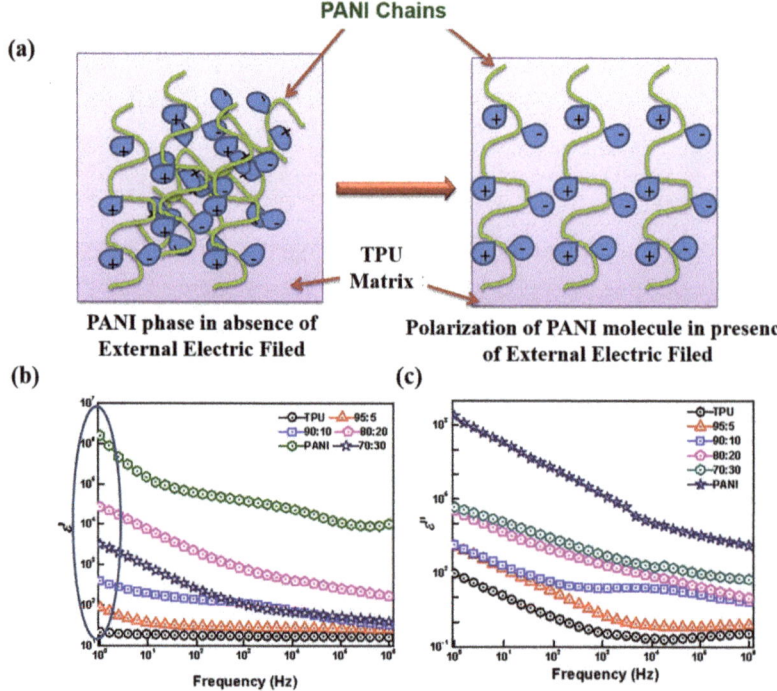

Figure 8. (a) Polarization of TPU/PANI composites under electric fields. The effect of PANI loading on ε' (b) and ε'' (c). Adapted with permission [94]. Copyright 2020, Springer Nature.

In our earlier study, CNCs were modified by methacrylic acid and then dispersed into UV curable resin methacrylate malate photocurable resin (MMPR) to prepare high dielectric CNCs-MAA/MMPR composites [93]. The modified CNCs can dispersed in the resin well and enhance the dielectric properties (Figure 9). With the addition of 1.0 wt% CNCs-MAA, the dielectric permittivity increased from 4.0 to 10.9 at 1 KHz, while the dielectric loss was only improved by 0.22. The influence of CNCs on the dielectric properties of composites mainly includes polar groups and the interface effect. The modified CNCs will have better dispersion in the matrix, but if the grafted polymer chain is too long, the interface effect will be affected, and the dielectric constant will be less improved. Asma Khouaja et al. used cellulose to enhance the dielectric permittivity of high density polyethylene (HDPE) [97]. With a 50% addition of cellulose, the dielectric permittivity of HDPE at 10^6 Hz increased from 1.3 to 2.2. This shows that it is feasible to enhance the dielectric properties of the composites by polymer fillers, such as CNCs, PANI, which provides a new insight for research on all-organic high dielectric polymer composites [98,99].

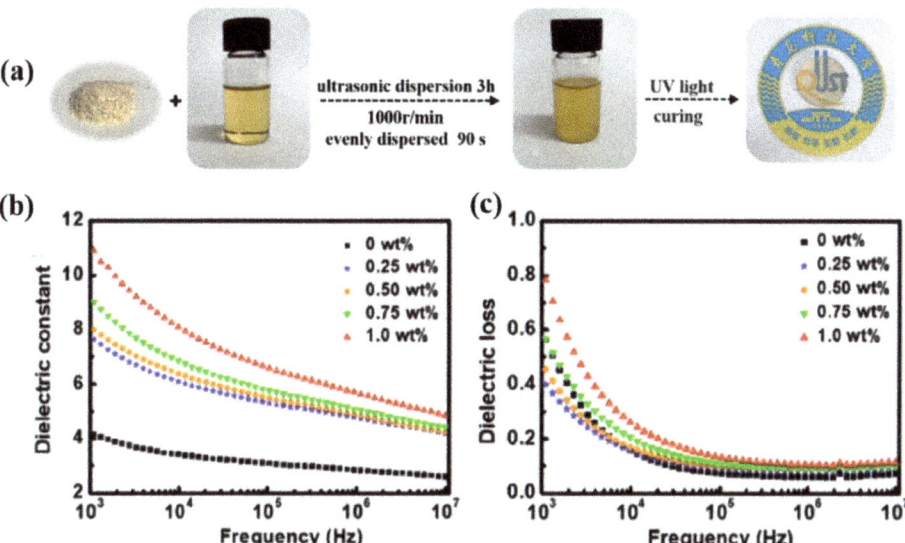

Figure 9. (**a**) preparation of MMPR/CNCs-MAA. Dielectric permittivity (**b**) and dielectric loss (**c**) of MMPR/CNCs-MAA. Adapted with permission [93]. Copyright 2022, Elsevier.

4. Influence of Structural Design to Dielectric Properties of Polymer Composites

The dielectric properties of composites are influenced not only by the matrix and filler but also by the distribution of the filler in the matrix. The orientation of the fillers in a certain direction can improve the dielectric permittivity in this direction to a large extent [46,55,57]. Shear force, magnetic fields, and electric fields have been developed to orient fillers in a specific direction to prepare dielectric materials of outstanding dielectric properties [100–103].

Under tensile or shear forces, fillers in the polymer matrix can be aligned in the direction of the force. Zhang et al. oriented $0.5Ba(Zr_{0.2}Ti_{0.8})O_3$-$0.5(Ba_{0.7}Ca_{0.3})TiO_3$ (BZCT) BZCT and BZCT@SiO_2 nanofibers in PVDF by shear force during electrospinning [104]. Two nanocomposites, BZCT-PVDF and BZCT@SiO_2-PVDF, were prepared by the hot-pressing method. As shown in Figure 10, the orientation direction of the nanofibers is perpendicular to the direction of the electric field and the orientation structure of fibers in the matrix is obvious. The dielectric permittivity of BZCT-PVDF at 10 Hz decreases from 24 to 22 when the nanofiber content is 15 vol% after orientation. Tang et al. used the uniaxial tensile method to orient lead zirconate titanate nanowires (PZT-NWs) in a thermoplastic elastomer [102]. The orientation of the oriented nanowires was also perpendicular to the electric field. At a frequency of 1 KHz with a PZT-NWs content of 40%, the dielectric permittivity of the oriented composite decreases from 40 to 25. The orientation can increase the amount of filler in a certain direction. If the orientation is wrong, the result will be a decrease in the amount of filler in this direction. Although the orientation of the force field can complete the orientation process smoothly and quickly, the orientation process often requires a large displacement, which greatly limits the orientation direction. Therefore, magnetic field and electric field, which are convenient and simple orientation methods, are favored by more researchers.

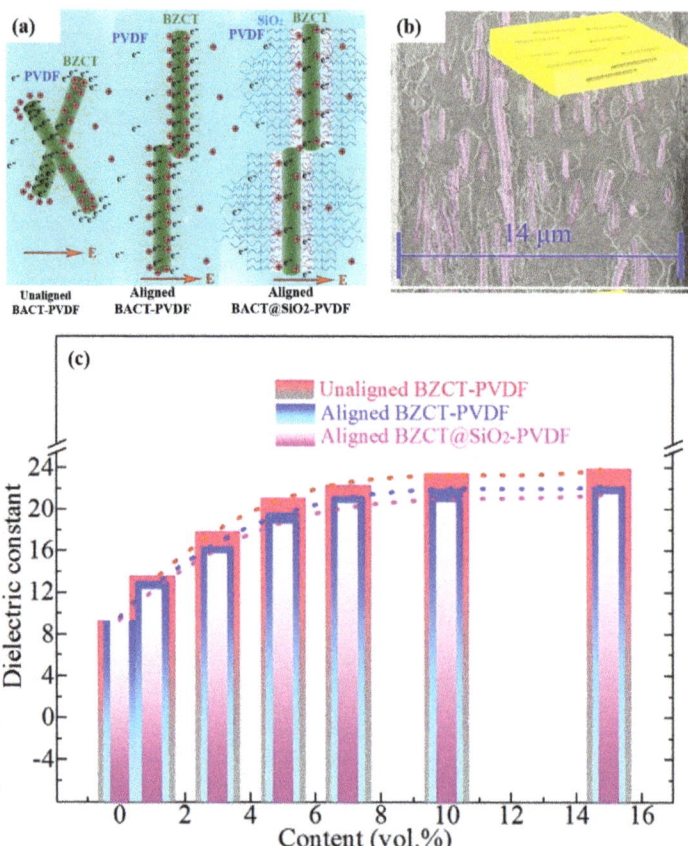

Figure 10. (**a**) Schematic illustration of polarization behavior of BZCT and BZCT@SiO$_2$ nanofibers. (**b**) SEM image of oriented structure. (**c**) Dielectric permittivity of composites after orientation at 10 Hz. Adapted with permission [104]. Copyright 2019, Elsevier.

The electric or magnetic fields can rapidly orient the fillers and produce thin films with oriented structures along the direction of electric fields. Chen et al. prepared high dielectric composites with excellent properties by orientations of silver-coated cellulose nanocrystals in silicone rubber by the electric field (Figure 11) [101]. The orientation process could be completed in only 90s. The dielectric permittivity of the oriented composites was significantly improved and the dielectric permittivity after orientation was increased from 13.8 to 38.6 at 10^{-2} Hz for 10 wt% silver plated CNC and the dielectric loss did not increase significantly. The silver-coated cellulose nanocrystals have a higher response rate under the electric field, which makes the cellulose nanocrystals more easily oriented. The orientation process increases the amount of cellulose in the direction of the electric filed, so the dielectric constant increases. Of course, at low filler content, the orientation process has little effect on the dielectric loss and if the content is high, the dielectric loss will rise sharply.

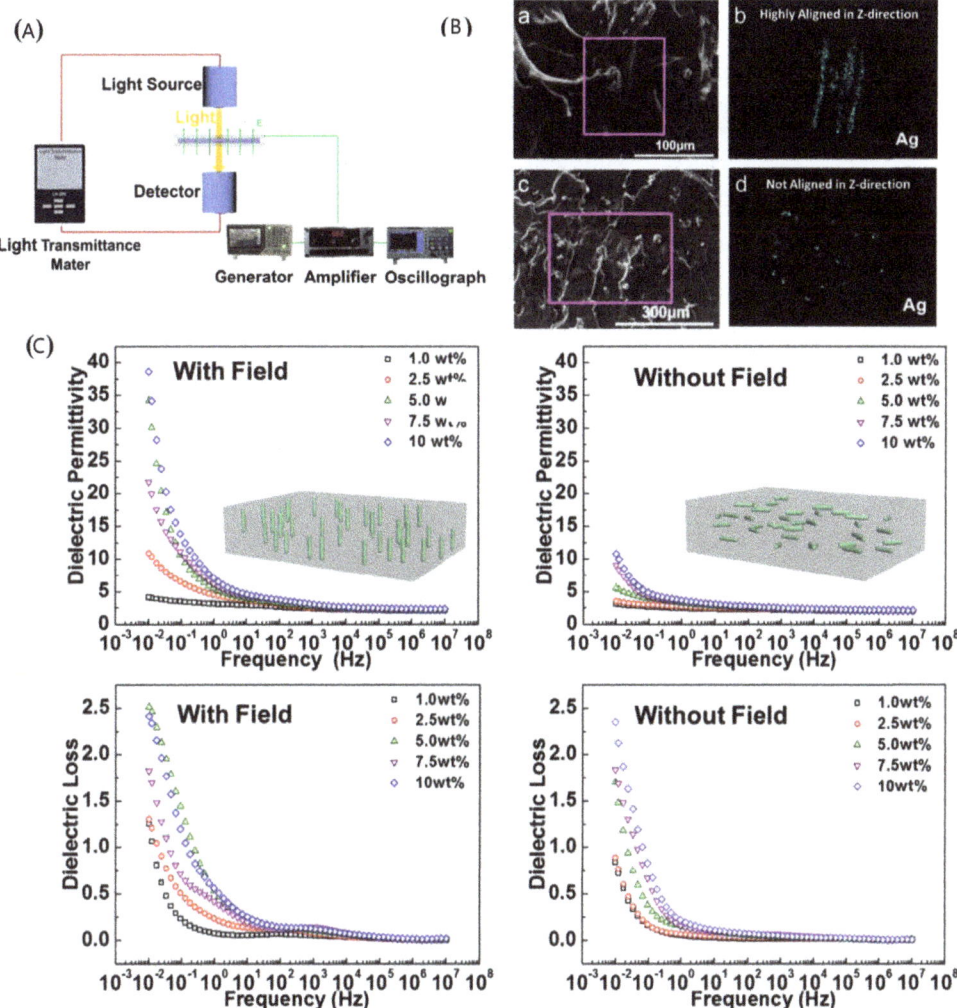

Figure 11. (**A**) Preparation of high dielectric composites by electric fields. (**B**) SEM image of oriented structures. (**C**) Dielectric properties of composites after orientation. Adapted with permission [101]. Copyright 2020, American Chemical Society.

5. Regulation Strategies of Dielectric Loss

High dielectric loss can seriously affect the practical application of high dielectric polymer composites. Dielectric loss represents the heat generated by the dielectric when it consumes part of the electric energy in the alternating electric field, which is mainly affected by polarization loss and conductivity loss. Polarization loss can hardly be avoided, while the conductivity loss caused dielectric loss can be effectively manipulated. Conductivity loss is caused by the current flow generated inside of the material. Therefore, conductance loss can be reduced by blocking the conductive pathways [105,106]. The key to reducing dielectric loss is to prevent the fillers from contacting each other to form conductive pathways, since the polymer matrix is not conductive. As Figure 12 shows, Wang et al. prepared rGO-PVA by polyvinyl alcohol modification of reduced grapheme oxide (rGO) to block the mutual contact between rGO in the PVDF matrix [107]. The dielectric permittivity of 2.2 vol% rGO-PVA/PVDF composite is up to 230 at 100 Hz and the dielectric loss remains low. While the dielectric loss of 2.2 vol% rGO/PVDF is as high as 50, which is unacceptable for practical energy storage use. It can be concluded that the coated insulation layers can effectively block the conductive pathways to suppress the rise of dielectric loss. The fillers are evenly dispersed into the matrix after being coated with insulation coating. In this way, the fillers can not only improve the dielectric constant of the composite but also prevent the filler from contacting each other, so as to achieve the purpose of inhibiting the dielectric loss. Therefore, in the report of wang et al., the dielectric constant of composite materials is increased while the dielectric loss is lower. Similarly, Yang et al. prepared PS/BaTiO$_3$ (BT-PS) by modifying BaTiO$_3$ with polystyrene (PS) and prepared composites with BT-PS as the filler and PS as the matrix [108]. The PS shell effectively inhibits the enhancement of dielectric loss. With addition of 47.69 vol% BaTiO$_3$, the dielectric loss of PS/BaTiO$_3$ composites at 1 KHZ was increased by as low as 0.005.

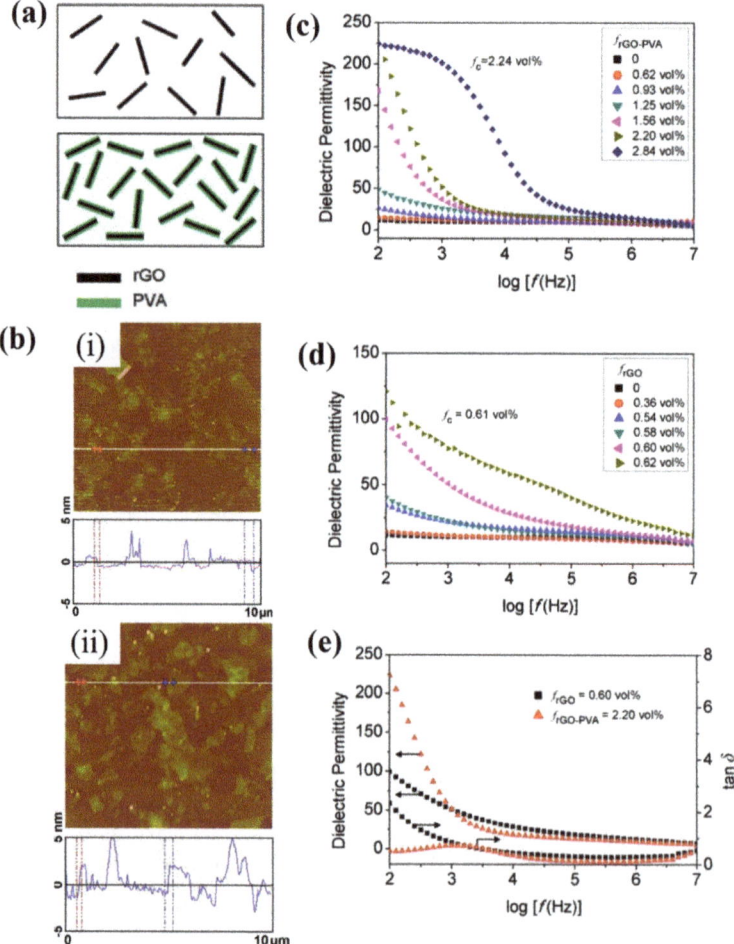

Figure 12. (**a**) Sketches of rGO and rGO-PVA. (**b**) AFM images of (**i**) rGO and (**ii**) rGO-PVA. Dielectric permittivity of (**c**) rGO-PVA/PVDF films and (**d**) rGO/PVDF films. (**e**) Dielectric properties of nanocomposites. Adapted with permission [107]. Copyright 2012, American Chemical Society.

In addition to the modification of fillers with polymers, incorporating insulative particles into potential conductive pathways is also an eye-catching approach. To prevent the formation of a conductive pathway, the method of using barium titanate ($BaTiO_3$) particles as a barrier to block the formation of a conductive pathway was ingeniously proposed and achieved quite good performance (Figure 13) [109]. As shown in Figure 13c,d, when the graphite content was 2.5 wt%, the dielectric loss increased to as high as 396 for aligned composites; however, the dielectric loss could be reduced to 0.19 by blocking the formation of the conductive pathway by adding 5 wt% insulating $BaTiO_3$. Note that the dielectric permittivity remained at a high value (73.5) after introducing insulating $BaTiO_3$. $BaTiO_3$ blocks the mutual contact of PANI and reduces the formation of conductive pathways, which can significantly reduce the conductivity loss in the composite, so the dielectric loss of the composite is still low after orientation.

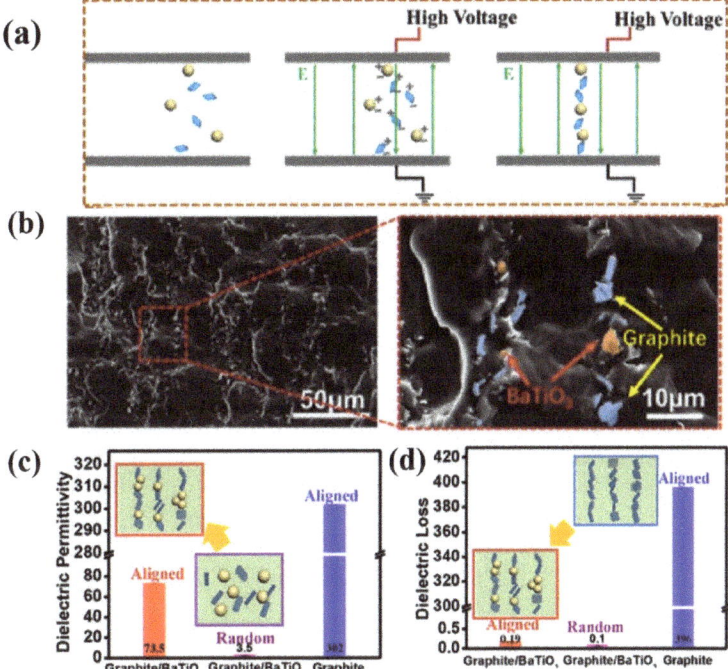

Figure 13. (a) A scheme of perparation of aligned copmposites. (b) Cross-sectional SEM images of the composites. Dielectric permittivity (c) and dielectric loss (d) of composite films. Adapted with permission [109]. Copyright 2021, Elsevier.

6. Summary and Outlook

In summary, the influences of the matrix, fillers, and filler arrangement on dielectric properties are systematically and comprehensively summarized, and the regulation strategies of dielectric loss are introduced as well. The effect of the polarity of the matrix on dielectric properties and the influences of conductive fillers, ceramic fillers, and polar polymer fillers on the dielectric properties of composites are described. The advantages and disadvantages of different type fillers are listed. The influence of dielectric properties of oriented structures and various orientation methods, including electric fields, shear force, and magnetic fields, are also introduced. Moreover, the methods to inhibit the increase of dielectric loss, including coating insulation and introducing insulation particles, are reviewed.

While considerable progress has been made in the research of high dielectric composites in recent years, challenges and bottlenecks still exists. Informed research and technologies in high dielectric composites are still limited to the laboratory state. The reported methods to inhibit the rise of dielectric loss are cumbersome, which limits its commercial process. In addition, the preparation of high dielectric composite materials is still dominated by traditional methods, which make it difficult to prepare complex and ingenious structures.

In the near future, the development of high dielectric polymer composites is no longer limited to the use of only one type of filler in one system. Instead, multiple fillers and multiple structures desiderate to be designed and combined in high dielectric polymer composites to enhance dielectric properties by synergistic effect [109,110]. New strategies of restraining dielectric loss are also urgently needed since the previous methods of coating insulating layer on fillers or introducing insulating fillers into conductive fillers are still difficult to scaleup in practical application. The application of biomaterials is also a new trend of high dielectric composites due to the growing piles of electronic trash. In order to solve the environmental problems caused by electronic trash, more and more biological matrix and bio-based fillers will be applied for the preparation of high dielectric composites. Note here, the long-term stability of the biological matrix and bio-based fillers during electronic environments need to be systematically investigated. Finally, 3D printing is needed in future studies on the preparation of high dielectric polymer composites, which will have positive implications for demonstrating the increasingly complicated electronic design in numerous fields.

Author Contributions: Conceptualization, Q.W.; software, Z.H.; validation, T.R., formal analysis, X.L.; investigation, J.C.; resources, J.Z.; data curation, Q.W.; writing—original draft preparation, Q.W.; writing—review and editing, Y.C.; visualization, W.W.; supervision, Y.C.; funding acquisition, X.L. All authors have read and agreed to the published version of the manuscript.

Funding: This work is supported by the National Natural Science Foundation of China (51803103) and the Program of National Key Research and Development of China (2022YFB3603702). The authors would also like to acknowledge financial support from the Opening Project of Key Laboratory of Optoelectronic Chemical Materials and Devices of Ministry of Education, Jianghan University (JDGD-202201) and Open Fund of Key Laboratory of Rubber Plastics, Ministry of Education/Shandong Provincial Key Laboratory of Rubber-plastics (KF2020002).

Institutional Review Board Statement: Not applicable.

Data Availability Statement: Not applicable.

Conflicts of Interest: The authors declare no conflict of interest.

References

1. Mendes-Felipe, C.; Barbosa, J.; Gonçalves, S.; Pereira, N.; Costa, C.; Vilas-Vilela, J.; Lanceros-Mendez, S. High dielectric constant UV curable polyurethane acrylate/indium tin oxide composites for capacitive sensing. *Compos. Sci. Technol.* **2020**, *199*, 108363. [CrossRef]
2. Asiri, A.M. Inamuddin. In *Ionic Polymer Metal Composites for Sensors and Actuators*; Springer: Berlin/Heidelberg, Germany, 2019. [CrossRef]
3. Zhang, Y.; Lin, B.; Sun, Y.; Han, P.; Wang, J.; Ding, X.; Zhang, X.; Yang, H. MoO2@ Cu@ C Composites Prepared by Using Polyoxometalates@ Metal-Organic Frameworks as Template for All-Solid-State Flexible Supercapacitor. *Electrochim. Acta* **2016**, *188*, 490–498. [CrossRef]
4. Ma, T.; Zhao, Y.; Ruan, K.; Liu, X.; Zhang, J.; Guo, Y.; Yang, X.; Kong, J.; Gu, J. Highly Thermal Conductivities, Excellent Mechanical Robustness and Flexibility, and Outstanding Thermal Stabilities of Aramid Nanofiber Composite Papers with Nacre-Mimetic Layered Structures. *ACS Appl. Mater. Interfaces* **2019**, *12*, 1677–1686. [CrossRef] [PubMed]
5. Zhang, D.-L.; Liu, S.-N.; Cai, H.-W.; Feng, Q.-K.; Zhong, S.-L.; Zha, J.-W.; Dang, Z.-M. Enhanced thermal conductivity and dielectric properties in electrostatic self-assembly 3D pBN@nCNTs fillers loaded in epoxy resin composites. *J. Materiomics* **2020**, *6*, 751–759. [CrossRef]
6. Thakur, V.K.; Gupta, R.K. Recent Progress on Ferroelectric Polymer-Based Nanocomposites for High Energy Density Capacitors: Synthesis, Dielectric Properties, and Future Aspects. *Chem. Rev.* **2016**, *116*, 4260–4317. [CrossRef]

7. Wang, B.; Huang, W.; Chi, L.; Al-Hashimi, M.; Marks, T.J.; Facchetti, A. High-*k* Gate Dielectrics for Emerging Flexible and Stretchable Electronics. *Chem. Rev.* **2018**, *118*, 5690–5754. [CrossRef]
8. Li, Y.; Krentz, T.M.; Wang, L.; Benicewicz, B.C.; Schadler, L.S. Ligand Engineering of Polymer Nanocomposites: From the Simple to the Complex. *ACS Appl. Mater. Interfaces* **2014**, *6*, 6005–6021. [CrossRef]
9. Yuan, J.-K.; Yao, S.-H.; Dang, Z.-M.; Sylvestre, A.; Genestoux, M.; Bai, J. Giant Dielectric Permittivity Nanocomposites: Realizing True Potential of Pristine Carbon Nanotubes in Polyvinylidene Fluoride Matrix through an Enhanced Interfacial Interaction. *J. Phys. Chem. C* **2011**, *115*, 5515–5521. [CrossRef]
10. Dang, Z.-M.; Yuan, J.-K.; Zha, J.-W.; Zhou, T.; Li, S.-T.; Hu, G.-H. Fundamentals, processes and applications of high-permittivity polymer–matrix composites. *Prog. Mater. Sci.* **2012**, *57*, 660–723. [CrossRef]
11. Fan, P.; Wang, L.; Yang, J.; Chen, F.; Zhong, M. Graphene/poly(vinylidene fluoride) composites with high dielectric constant and low percolation threshold. *Nanotechnology* **2012**, *23*, 365702. [CrossRef]
12. Feng, Q.-K.; Zhong, S.-L.; Pei, J.-Y.; Zhao, Y.; Zhang, D.-L.; Liu, D.-F.; Zhang, Y.-X.; Dang, Z.-M. Recent Progress and Future Prospects on All-Organic Polymer Dielectrics for Energy Storage Capacitors. *Chem. Rev.* **2021**, *122*, 3820–3878. [CrossRef] [PubMed]
13. Patel, P.K.; Rani, J.; Yadav, K. Effective strategies for reduced dielectric loss in ceramic/polymer nanocomposite film. *Ceram. Int.* **2020**, *47*, 10096–10103. [CrossRef]
14. Aepuru, R.; Mondal, S.; Ghorai, N.; Kumar, V.; Panda, H.S.; Ghosh, H.N. Exploring the Carrier Dynamics in Zinc Oxide–Metal Halide-Based Perovskite Nanostructures: Toward Reduced Dielectric Loss and Improved Photocurrent. *J. Phys. Chem. C* **2018**, *122*, 27273–27283. [CrossRef]
15. Li, Q.; Cheng, J.; Chen, J. Reduced dielectric loss and enhanced piezoelectric properties of Mn modified 0.71 $BiFeO_3$–0.29 $BaTiO_3$ ceramics sintered under oxygen atmosphere. *J. Mater. Sci. Mater. Electron.* **2017**, *28*, 1370–1377. [CrossRef]
16. Deng, Y.-Y.; Zhou, D.-L.; Han, D.; Zhang, Q.; Chen, F.; Fu, Q. Fluoride ion encapsulated polyhedral oligomeric silsesquioxane: A novel filler for polymer nanocomposites with enhanced dielectric constant and reduced dielectric loss. *Compos. Sci. Technol.* **2020**, *189*, 108035. [CrossRef]
17. Hu, Z.; Wang, Y.; Liu, X.; Wang, Q.; Cui, X.; Jin, S.; Yang, B.; Xia, Y.; Huang, S.; Qiang, Z.; et al. Rational design of POSS containing low dielectric resin for SLA printing electronic circuit plate composites. *Compos. Sci. Technol.* **2022**, *223*, 109403. [CrossRef]
18. Zhang, Z.; Tian, D.; Niu, Z.; Zhou, Y.; Hou, X.; Ma, X. Enhanced toughness and lowered dielectric loss of reactive POSS modified bismaleimide resin as well as the silica fiber reinforced composites. *Polym. Compos.* **2021**, *42*, 6900–6911. [CrossRef]
19. Zhu, L. Exploring Strategies for High Dielectric Constant and Low Loss Polymer Dielectrics. *J. Phys. Chem. Lett.* **2014**, *5*, 3677–3687. [CrossRef]
20. Zhu, L.; Wang, Q. Novel Ferroelectric Polymers for High Energy Density and Low Loss Dielectrics. *Macromolecules* **2012**, *45*, 2937–2954. [CrossRef]
21. He, D.; Wang, Y.; Song, S.; Liu, S.; Luo, Y.; Deng, Y. Polymer-based nanocomposites employing Bi_2S_3@SiO_2 nanorods for high dielectric performance: Understanding the role of interfacial polarization in semiconductor-insulator core-shell nanostructure. *Compos. Sci. Technol.* **2017**, *151*, 25–33. [CrossRef]
22. Marx, P.; Wanner, A.J.; Zhang, Z.; Jin, H.; Tsekmes, I.-A.; Smit, J.J.; Kern, W.; Wiesbrock, F. Effect of Interfacial Polarization and Water Absorption on the Dielectric Properties of Epoxy-Nanocomposites. *Polymers* **2017**, *9*, 195. [CrossRef] [PubMed]
23. Wang, Q.; Wu, C.; LaChance, A.M.; Zhou, J.; Gao, Y.; Zhang, Y.; Sun, L.; Cao, Y.; Liang, X. Interfacial polarization suppression of P(VDF-HFP) film through 2D montmorillonite nanosheets coating. *Prog. Org. Coat.* **2022**, *172*, 107119. [CrossRef]
24. Zhang, X.; Ye, H.; Xu, L. Exploring the interfacial polarization in poly(vinylidene fluoride-chlorotrifluoroethylene) dielectric film with regulated surface conductivity of C@BT particles. *Appl. Surf. Sci.* **2022**, *600*, 154113. [CrossRef]
25. Huang, X.; Jiang, P.; Tanaka, T. A review of dielectric polymer composites with high thermal conductivity. *IEEE Electr. Insul. Mag.* **2011**, *27*, 8–16. [CrossRef]
26. Han, C.; Zhang, X.; Chen, D.; Ma, Y.; Zhao, C.; Yang, W. Enhanced dielectric properties of sandwich-structured biaxially oriented polypropylene by grafting hyper-branched aromatic polyamide as surface layers. *J. Appl. Polym. Sci.* **2020**, *137*, 48990. [CrossRef]
27. Mohamed, A.T. Experimental enhancement for dielectric strength of polyethylene insulation materials using cost-fewer nanoparticles. *Int. J. Electr. Power Energy Syst.* **2015**, *64*, 469–475. [CrossRef]
28. Barber, P.; Balasubramanian, S.; Anguchamy, Y.; Gong, S.; Wibowo, A.; Gao, H.; Ploehn, H.J.; Zur Loye, H.-C. Polymer Composite and Nanocomposite Dielectric Materials for Pulse Power Energy Storage. *Materials* **2009**, *2*, 1697–1733. [CrossRef]
29. Molberg, M.; Crespy, D.; Rupper, P.; Nüesch, F.; Månson, J.-A.E.; Löwe, C.; Opris, D.M. High Breakdown Field Dielectric Elastomer Actuators Using Encapsulated Polyaniline as High Dielectric Constant Filler. *Adv. Funct. Mater.* **2010**, *20*, 3280–3291. [CrossRef]
30. Abouhaswa, A.S.; Taha, T.A. Tailoring the optical and dielectric properties of PVC/CuO nanocomposites. *Polym. Bull.* **2019**, *77*, 6005–6016. [CrossRef]
31. Pan, J.; Li, K.; Chuayprakong, S.; Hsu, T.; Wang, Q. High-Temperature Poly(phthalazinone ether ketone) Thin Films for Dielectric Energy Storage. *ACS Appl. Mater. Interfaces* **2010**, *2*, 1286–1289. [CrossRef]
32. Coburn, J.C.; Boyd, R.H. Dielectric relaxation in poly(ethylene terephthalate). *Macromolecules* **1986**, *19*, 2238–2245. [CrossRef]
33. Goyal, R.K.; Madav, V.V.; Pakankar, P.R.; Butee, S.P. Fabrication and properties of novel polyetheretherketone/barium ti-tanate composites with low dielectric loss. *J. Electron. Mater.* **2011**, *40*, 2240–2247. [CrossRef]

34. Pan, J.; Li, K.; Li, J.; Hsu, T.; Wang, Q. Dielectric characteristics of poly(ether ketone ketone) for high temperature capacitive energy storage. *Appl. Phys. Lett.* **2009**, *95*, 022902. [CrossRef]
35. Fang, L.; Wu, C.; Qian, R.; Xie, L.; Yang, K.; Jiang, P. Nano–micro structure of functionalized boron nitride and aluminum oxide for epoxy composites with enhanced thermal conductivity and breakdown strength. *RSC Adv.* **2014**, *4*, 21010–21017. [CrossRef]
36. Song, Y.; Shen, Y.; Liu, H.; Lin, Y.; Li, M.; Nan, C.-W. Improving the dielectric constants and breakdown strength of polymer composites: Effects of the shape of the BaTiO$_3$ nanoinclusions, surface modification and polymer matrix. *J. Mater. Chem.* **2012**, *22*, 16491–16498. [CrossRef]
37. Paniagua, S.A.; Kim, Y.; Henry, K.; Kumar, R.; Perry, J.W.; Marder, S.R. Surface-Initiated Polymerization from Barium Titanate Nanoparticles for Hybrid Dielectric Capacitors. *ACS Appl. Mater. Interfaces* **2014**, *6*, 3477–3482. [CrossRef]
38. Xie, L.; Huang, X.; Huang, Y.; Yang, K.; Jiang, P. Core@Double-Shell Structured BaTiO$_3$–Polymer Nanocomposites with High Dielectric Constant and Low Dielectric Loss for Energy Storage Application. *J. Phys. Chem. C* **2013**, *117*, 22525–22537. [CrossRef]
39. Gross, S.; Camozzo, D.; DI Noto, V.; Armelao, L.; Tondello, E. PMMA: A key macromolecular component for dielectric low-κ hybrid inorganic–organic polymer films. *Eur. Polym. J.* **2007**, *43*, 673–696. [CrossRef]
40. Chen, T.; Zhao, Y.; Pan, L.; Lin, M. Insight into effect of hydrothermal preparation process of nanofillers on dielectric, creep and electromechanical performance of polyurethane dielectric elastomer/reduced graphene oxide composites. *J. Mater. Sci. Mater. Electron.* **2015**, *26*, 10164–10171. [CrossRef]
41. Das, A.; Sinha, S.; Mukherjee, A.; Meikap, A. Enhanced dielectric properties in polyvinyl alcohol—Multiwall carbon nanotube composites. *Mater. Chem. Phys.* **2015**, *167*, 286–294. [CrossRef]
42. Tuncer, E.; Sauers, I.; James, D.R.; Ellis, A.R.; Duckworth, R.C. Nanodielectric system for cryogenic applications: Barium titanate filled polyvinyl alcohol. *IEEE Trans. Dielectr. Electr. Insul.* **2008**, *15*, 236–242. [CrossRef]
43. Shen, Y.; Lin, Y.H.; Nan, C.-W. Interfacial Effect on Dielectric Properties of Polymer Nanocomposites Filled with Core/Shell-Structured Particles. *Adv. Funct. Mater.* **2007**, *17*, 2405–2410. [CrossRef]
44. Xie, X.; Yang, C.; Qi, X.-D.; Yang, J.-H.; Zhou, Z.-W.; Wang, Y. Constructing polymeric interlayer with dual effects toward high dielectric constant and low dielectric loss. *Chem. Eng. J.* **2019**, *366*, 378–389. [CrossRef]
45. Gong, Y.; Zhou, W.; Sui, X.; Kou, Y.; Xu, L.; Duan, Y.; Chen, F.; Li, Y.; Liu, X.; Cai, H.; et al. Core-shell structured Al/PVDF nanocomposites with high dielectric permittivity but low loss and enhanced thermal conductivity. *Polym. Eng. Sci.* **2018**, *59*, 103–111. [CrossRef]
46. Zhou, W.; Chen, Q.; Sui, X.; Dong, L.; Wang, Z. Enhanced thermal conductivity and dielectric properties of Al/β-SiCw/PVDF composites. *Compos. Part A Appl. Sci. Manuf.* **2015**, *71*, 184–191. [CrossRef]
47. Salehiyan, R.; Nofar, M.; Makwakwa, D.; Ray, S.S. Shear-Induced Carbon Nanotube Migration and Morphological Development in Polylactide/Poly(vinylidene fluoride) Blend Nanocomposites and Their Impact on Dielectric Constants and Rheological Properties. *J. Phys. Chem. C* **2020**, *124*, 9536–9547. [CrossRef]
48. Zhang, L.; Shan, X.; Bass, P.; Tong, Y.; Rolin, T.D.; Hill, C.W.; Brewer, J.C.; Tucker, D.S.; Cheng, Z.-Y. Process and Microstructure to Achieve Ultra-high Dielectric Constant in Ceramic-Polymer Composites. *Sci. Rep.* **2016**, *6*, 35763. [CrossRef]
49. Dai, Z.-H.; Li, T.; Gao, Y.; Xu, J.; He, J.; Weng, Y.-X.; Guo, B.-H. Achieving high dielectric permittivity, high breakdown strength and high efficiency by cross-linking of poly(vinylidene fluoride)/BaTiO$_3$ nanocomposites. *Compos. Sci. Technol.* **2018**, *169*, 142–150. [CrossRef]
50. Liu, W.; Lee, S.W.; Lin, D.; Shi, F.; Wang, S.; Sendek, A.D.; Cui, Y. Enhancing ionic conductivity in composite polymer electrolytes with well-aligned ceramic nanowires. *Nat. Energy* **2017**, *2*, 17035. [CrossRef]
51. Zhang, Q.M.; Li, H.; Poh, M.; Xia, F.; Cheng, Z.-Y.; Xu, H.; Huang, C. An all-organic composite actuator material with a high dielectric constant. *Nature* **2002**, *419*, 284–287. [CrossRef]
52. Ltaief, A.O.; Ghorbel, N.; Benhamou, K.; Arous, M.; Kaddami, H.; Kallel, A. Impact of cellulose nanocrystals reinforcement on molecular dynamics and dielectric properties of PCL-based polyurethane. *Polym. Compos.* **2021**, *42*, 2737–2750. [CrossRef]
53. Ladhar, A.; Ben Mabrouk, A.; Arous, M.; Boufi, S.; Kallel, A. Dielectric properties of nanocomposites based on cellulose nanocrystals (CNCs) and poly(styrene-co-2-ethyl hexylacrylate) copolymer. *Polymer* **2017**, *125*, 76–89. [CrossRef]
54. Ma, L.; Liu, R.; Niu, H.; Zhao, M.; Huang, Y. Flexible and freestanding electrode based on polypyrrole/graphene/bacterial cellulose paper for supercapacitor. *Compos. Sci. Technol.* **2016**, *137*, 87–93. [CrossRef]
55. Guo, Y.; Chen, Y.; Wang, E.; Cakmak, M. Roll-to-Roll Continuous Manufacturing Multifunctional Nanocomposites by Electric-Field-Assisted "Z" Direction Alignment of Graphite Flakes in Poly(dimethylsiloxane). *ACS Appl. Mater. Interfaces* **2017**, *9*, 919–929. [CrossRef] [PubMed]
56. Martin, J.J.; Fiore, B.E.; Erb, R.M. Designing bioinspired composite reinforcement architectures via 3D magnetic printing. *Nat. Commun.* **2015**, *6*, 8641. [CrossRef] [PubMed]
57. Xu, S.; Liu, D.; Zhang, Q.; Fu, Q. Electric field-induced alignment of nanofibrillated cellulose in thermoplastic polyurethane matrix. *Compos. Sci. Technol.* **2017**, *156*, 117–126. [CrossRef]
58. Kadimi, A.; Benhamou, K.; Ounaies, Z.; Magnin, A.; Dufresne, A.; Kaddami, H.; Raihane, M. Electric Field Alignment of Nanofibrillated Cellulose (NFC) in Silicone Oil: Impact on Electrical Properties. *ACS Appl. Mater. Interfaces* **2014**, *6*, 9418–9425. [CrossRef]

59. Tsyganov, A.; Vikulova, M.; Artyukhov, D.; Bainyashev, A.; Goffman, V.; Gorokhovsky, A.; Boychenko, E.; Burmistrov, I.; Gorshkov, N. Permittivity and Dielectric Loss Balance of PVDF/$K_{1.6}Fe_{1.6}Ti_{6.4}O_{16}$/MWCNT Three-Phase Composites. *Polymers* **2022**, *14*, 4609. [CrossRef]
60. Deeba, F.; Gupta, A.K.; Kulshrestha, V.; Bafna, M.; Jain, A. Analysing the dielectric properties of ZnO doped PVDF/PMMA blend composite. *J. Mater. Sci. Mater. Electron.* **2022**, *33*, 23703–23713. [CrossRef]
61. Celebi, H.; Duran, S.; Dogan, A. The effect of core-shell $BaTiO_3$@ $SiO2$ on the mechanical and dielectric properties of PVDF composites. *Polym.-Plast. Technol. Mater.* **2022**, *61*, 1191–1203. [CrossRef]
62. Silakaew, K.; Swatsitang, E.; Thongbai, P. Novel polymer composites of RuO_2@ $nBaTiO_3$/PVDF with a high dielectric constant. *Ceram. Int.* **2022**, *48*, 18925–18932. [CrossRef]
63. Jing, L.; Li, W.; Gao, C.; Li, M.; Fei, W. Excellent energy storage properties achieved in PVDF-based composites by designing the lamellar-structured fillers. *Compos. Sci. Technol.* **2022**, *227*, 109568. [CrossRef]
64. Zhang, T.; Sun, Q.; Kang, F.; Wang, Z.; Xue, R.; Wang, J.; Zhang, L. Sandwich-structured polymer dielectric composite films for improving breakdown strength and energy density at high temperature. *Compos. Sci. Technol.* **2022**, *227*, 109596. [CrossRef]
65. Li, Z.; Shen, Z.; Yang, X.; Zhu, X.; Zhou, Y.; Dong, L.; Xiong, C.; Wang, Q. Ultrahigh charge-discharge efficiency and enhanced energy density of the sandwiched polymer nanocomposites with poly(methyl methacrylate) layer. *Compos. Sci. Technol.* **2020**, *202*, 108591. [CrossRef]
66. Liao, Y.; Weng, Y.; Wang, J.; Zhou, H.; Lin, J.; He, S. Silicone Rubber Composites with High Breakdown Strength and Low Dielectric Loss Based on Polydopamine Coated Mica. *Polymers* **2019**, *11*, 2030. [CrossRef]
67. Anjeline, C.J.; Mali, D.; Lakshminarasimhan, N. High dielectric constant of $NiFe_2O_4$–$LaFeO_3$ nanocomposite: Interfacial conduction and dielectric loss. *Ceram. Int.* **2021**, *47*, 34278–34288. [CrossRef]
68. Wang, H.; Wang, Q.; Zhang, Q.; Yang, H.; Dong, J.; Cheng, J.; Tong, J.; Wen, J. High thermal conductive composite with low dielectric constant and dielectric loss accomplished through flower-like Al_2O_3 coated BNNs for advanced circuit substrate applications. *Compos. Sci. Technol.* **2021**, *216*, 109048. [CrossRef]
69. Huang, Z.-X.; Zhao, M.-L.; Zhang, G.-Z.; Song, J.; Qu, J.-P. Controlled localizing multi-wall carbon nanotubes in polyvinylidene fluoride/acrylonitrile butadiene styrene blends to achieve balanced dielectric constant and dielectric loss. *Compos. Sci. Technol.* **2021**, *212*, 108874. [CrossRef]
70. Ding, X.; Pan, Z.; Zhang, Y.; Shi, S.; Cheng, Y.; Chen, H.; Li, Z.; Fan, X.; Liu, J.; Yu, J.; et al. Regulation of Interfacial Polarization and Local Electric Field Strength Achieved Highly Energy Storage Performance in Polyetherimide Nanocomposites at Elevated Temperature via 2D Hybrid Structure. *Adv. Mater. Interfaces* **2022**, *9*, 2201100. [CrossRef]
71. Bouiri, E.M.; Farhan, R.; Chakhchaoui, N.; Oumghar, K.; Denktas, C.; Eddiai, A.; Meddad, M.; Mazroui, M.; Cherkaoui, O.; Omari, L.E.H. Improving dielectric properties of composites thin films with polylactic acid and PZT microparticles induced by interfacial polarization. *Eur. Phys. J. Appl. Phys.* **2022**, *97*, 64. [CrossRef]
72. Bronnikov, S.; Kostromin, S.; Asandulesa, M.; Pankin, D.; Podshivalov, A. Interfacial interactions and interfacial polarization in polyazomethine/MWCNTs nanocomposites. *Compos. Sci. Technol.* **2020**, *190*, 108049. [CrossRef]
73. Kuang, X.; Liu, Z.; Zhu, H. Dielectric properties of Ag@ C/PVDF composites. *J. Appl. Polym. Sci.* **2013**, *129*, 3411–3416. [CrossRef]
74. Tuichai, W.; Kum-Onsa, P.; Danwittayakul, S.; Manyam, J.; Harnchana, V.; Thongbai, P.; Phromviyo, N.; Chindaprasirt, P. Significantly Enhanced Dielectric Properties of Ag-Deposited $(In_{1/2}Nb_{1/2})_{0.1}Ti_{0.9}O_2$/PVDF Polymer Composites. *Polymers* **2021**, *13*, 1788. [CrossRef] [PubMed]
75. Kaur, M.; Kumar, V.; Singh, J.; Datt, J.; Sharma, R. Effect of Cu-N co-doping on the dielectric properties of ZnO nanoparticles. *Mater. Technol.* **2022**, *37*, 2644–2658. [CrossRef]
76. Huang, A.; Liu, F.; Cui, Z.; Wang, H.; Song, X.; Geng, L.; Peng, X. Novel PTFE/CNT composite nanofiber membranes with enhanced mechanical, crystalline, conductive, and dielectric properties fabricated by emulsion electrospinning and sintering. *Compos. Sci. Technol.* **2021**, *214*, 108980. [CrossRef]
77. Fan, X.; Zhang, A.; Li, M.; Xu, H.; Xue, J.; Ye, F.; Cheng, L. A reduced graphene oxide/bi-MOF-derived carbon composite as high-performance microwave absorber with tunable dielectric properties. *J. Mater. Sci. Mater. Electron.* **2020**, *31*, 11774–11783. [CrossRef]
78. Zhang, L.; Yuan, S.; Chen, S.; Wang, D.; Han, B.-Z.; Dang, Z.-M. Preparation and dielectric properties of core–shell structured Ag@polydopamine/poly(vinylidene fluoride) composites. *Compos. Sci. Technol.* **2015**, *110*, 126–131. [CrossRef]
79. Wang, Y.; Zhu, L.; Zhou, J.; Jia, B.; Jiang, Y.; Wang, J.; Wang, M.; Cheng, Y.; Wu, K. Dielectric properties and thermal conductivity of epoxy resin composite modified by Zn/ZnO/Al2O3 core–shell particles. *Polym. Bull.* **2018**, *76*, 3957–3970. [CrossRef]
80. Mei, B.; Qin, Y.; Agbolaghi, S. A review on supramolecules/nanocomposites based on carbonic precursors and dielectric/conductive polymers and their applications. *Mater. Sci. Eng. B* **2021**, *269*, 115181. [CrossRef]
81. Chen, Y.; Lin, B.; Zhang, X.; Wang, J.; Lai, C.; Sun, Y.; Liu, Y.; Yang, H. Enhanced dielectric properties of amino-modified-CNT/polyimide composite films with a sandwich structure. *J. Mater. Chem. A* **2014**, *2*, 14118–14126. [CrossRef]
82. Wang, J. High-Performance Dielectric Ceramic for Energy Storage Capacitors. *Coatings* **2022**, *12*, 889. [CrossRef]
83. Feng, Y.; Li, W.L.; Hou, Y.F.; Yu, Y.; Cao, W.P.; Zhang, T.D.; Fei, W.D. Enhanced dielectric properties of PVDF-HFP/$BaTiO_3$-nanowire composites induced by interfacial polarization and wire-shape. *J. Mater. Chem. C* **2014**, *3*, 1250–1260. [CrossRef]
84. Pan, Z.; Yao, L.; Zhai, J.; Shen, B.; Wang, H. Significantly improved dielectric properties and energy density of polymer nanocomposites via small loaded of $BaTiO_3$ nanotubes. *Compos. Sci. Technol.* **2017**, *147*, 30–38. [CrossRef]

85. Ding, C.; Yu, S.; Tang, X.; Liu, Z.; Luo, H.; Zhang, Y.; Zhang, D.; Chen, S. The design and preparation of high-performance ABS-based dielectric composites via introducing core-shell polar polymers@ BaTiO$_3$ nanoparticles. *Compos. Part A Appl. Sci. Manuf.* **2022**, *163*, 107214. [CrossRef]
86. Costa, C.; Reizabal, A.; i Serra, R.S.; Balado, A.A.; Pérez-Álvarez, L.; Ribelles, J.G.; Vilas-Vilela, J.; Lanceros-Méndez, S. Broadband dielectric response of silk Fibroin/BaTiO$_3$ composites: Influence of nanoparticle size and concentration. *Compos. Sci. Technol.* **2021**, *213*, 108927. [CrossRef]
87. Chen, J.; Huang, F.; Zhang, C.; Meng, F.; Cao, L.; Lin, H. Enhanced energy storage density in poly(vinylidene fluoride-hexafluoropropylene) nanocomposites by filling with core-shell structured BaTiO$_3$@ MgO nanoparticals. *J. Energy Storage* **2022**, *53*, 105163. [CrossRef]
88. Chen, J.; Zhou, C.; Cai, W.; Huang, F.; Zhang, C.; Cao, L.; Meng, F. Pluronic F127-modified BaTiO$_3$ for ceramic/polymer nanocomposite dielectric capacitor with enhanced energy storage performance. *Polym. Eng. Sci.* **2022**, *62*, 1811–1822. [CrossRef]
89. Li, H.; Fu, Y.; Alhashmialameer, D.; Thabet, H.K.; Zhang, P.; Wang, C.; Zhu, K.; Huang, M.; Guo, Z.; Dang, F. Lattice distortion embedded core–shell nanoparticle through epitaxial growth barium titanate shell on the strontium titanate core with enhanced dielectric response. *Adv. Compos. Hybrid Mater.* **2022**, *5*, 2631–2641. [CrossRef]
90. Yang, K.; Huang, X.; Zhu, M.; Xie, L.; Tanaka, T.; Jiang, P. Combining RAFT Polymerization and Thiol–Ene Click Reaction for Core–Shell Structured Polymer@BaTiO$_3$ Nanodielectrics with High Dielectric Constant, Low Dielectric Loss, and High Energy Storage Capability. *ACS Appl. Mater. Interfaces* **2014**, *6*, 1812–1822. [CrossRef]
91. Wu, W.; Ren, T.; Liu, X.; Huai, K.; Cui, X.; Wei, H.; Hu, J.; Xia, Y.; Huang, S.; Fu, K.; et al. Electric field-assisted preparation of PANI/TPU all-organic composites with enhanced dielectric permittivity and anisotropic optical properties. *Polym. Eng. Sci.* **2022**, *62*, 3945–3951. [CrossRef]
92. Wei, H.; Yuan, Y.; Ren, T.; Zhou, L.; Liu, X.; Saeed, H.A.M.; Jin, P.; Chen, Y. High-Dielectric PVP@PANI/PDMS Composites Fabricated via an Electric Field-Assisted Approach. *Polymers* **2022**, *14*, 4381. [CrossRef] [PubMed]
93. Wang, Q.; Liu, X.; Qiang, Z.; Hu, Z.; Cui, X.; Wei, H.; Hu, J.; Xia, Y.; Huang, S.; Zhang, J.; et al. Cellulose nanocrystal enhanced, high dielectric 3D printing composite resin for energy applications. *Compos. Sci. Technol.* **2022**, *227*, 109601. [CrossRef]
94. Dash, K.; Hota, N.K.; Sahoo, B.P. Fabrication of thermoplastic polyurethane and polyaniline conductive blend with improved mechanical, thermal and excellent dielectric properties: Exploring the effect of ultralow-level loading of SWCNT and temperature. *J. Mater. Sci.* **2020**, *55*, 12568–12591. [CrossRef]
95. Zhang, Q.; Jiang, Y.; Yu, E.; Yang, H. Significantly enhanced dielectric properties of P(VDF-HFP) composite films filled with core-shell BaTiO$_3$@ PANI nanoparticles. *Surf. Coat. Technol.* **2018**, *358*, 293–298. [CrossRef]
96. Rahnamol, A.M.; Gopalakrishnan, J. Improved dielectric and dynamic mechanical properties of epoxy/polyaniline nanorod/ *in situ* reduced graphene oxide hybrid nanocomposites. *Polym. Compos.* **2020**, *41*, 2998–3013. [CrossRef]
97. Khouaja, A.; Koubaa, A.; Ben Daly, H. Dielectric properties and thermal stability of cellulose high-density polyethylene bio-based composites. *Ind. Crop. Prod.* **2021**, *171*, 113928. [CrossRef]
98. Guo, Q.; Xue, Q.; Wu, T.; Pan, X.; Zhang, J.; Li, X.; Zhu, L. Excellent dielectric properties of PVDF-based composites filled with carbonized PAN/PEG copolymer fibers. *Compos. Part A Appl. Sci. Manuf.* **2016**, *87*, 46–53. [CrossRef]
99. Wang, P.; Yin, Y.; Fang, L.; He, J.; Wang, Y.; Cai, H.; Yang, Q.; Shi, Z.; Xiong, C. Flexible cellulose/PVDF composite films with improved breakdown strength and energy density for dielectric capacitors. *Compos. Part A Appl. Sci. Manuf.* **2023**, *164*, 107325. [CrossRef]
100. Chen, Y.; Liu, Y.; Yang, J.; Zhang, B.; Hu, Z.; Wang, Q.; Wu, W.; Shang, Y.; Xia, Y.; Duan, Y.; et al. Fabrication of high dielectric permittivity polymer composites by architecting aligned micro-enhanced-zones of ultralow content graphene using electric fields. *Mater. Today Commun.* **2019**, *21*, 100649. [CrossRef]
101. Chen, Y.; Liu, Y.; Xia, Y.; Liu, X.; Qiang, Z.; Yang, J.; Zhang, B.; Hu, Z.; Wang, Q.; Wu, W.; et al. Electric Field-Induced Assembly and Alignment of Silver-Coated Cellulose for Polymer Composite Films with Enhanced Dielectric Permittivity and Anisotropic Light Transmission. *ACS Appl. Mater. Interfaces* **2020**, *12*, 24242–24249. [CrossRef]
102. Tang, H.; Lin, Y.; Sodano, H.A. Enhanced Energy Storage in Nanocomposite Capacitors through Aligned PZT Nanowires by Uniaxial Strain Assembly. *Adv. Energy Mater.* **2012**, *2*, 469–476. [CrossRef]
103. Zhang, X.; Jiang, J.; Shen, Z.; Dan, Z.; Li, M.; Lin, Y.; Nan, C.; Chen, L.; Shen, Y. Polymer Nanocomposites with Ultrahigh Energy Density and High Discharge Efficiency by Modulating their Nanostructures in Three Dimensions. *Adv. Mater.* **2018**, *30*, e1707269. [CrossRef] [PubMed]
104. Zhang, Y.; Zhang, C.; Feng, Y.; Zhang, T.; Chen, Q.; Chi, Q.; Liu, L.; Li, G.; Cui, Y.; Wang, X. Excellent energy storage performance and thermal property of polymer-based composite induced by multifunctional one-dimensional nanofibers oriented in-plane direction. *Nano Energy* **2019**, *56*, 138–150. [CrossRef]
105. Luo, H.; Zhou, X.; Ellingford, C.; Zhang, Y.; Chen, S.; Zhou, K.; Zhang, D.; Bowen, C.R.; Wan, C. Interface design for high energy density polymer nanocomposites. *Chem. Soc. Rev.* **2019**, *48*, 4424–4465. [CrossRef] [PubMed]
106. Huang, Y.; Huang, X.; Schadler, L.S.; He, J.; Jiang, P. Core@ Double-Shell Structured Nanocomposites: A Route to High Dielectric Constant and Low Loss Material. *ACS Appl. Mater. Interfaces* **2016**, *8*, 25496–25507. [CrossRef] [PubMed]
107. Wang, D.; Bao, Y.; Zha, J.-W.; Zhao, J.; Dang, Z.-M.; Hu, G.-H. Improved Dielectric Properties of Nanocomposites Based on Poly(vinylidene fluoride) and Poly(vinyl alcohol)-Functionalized Graphene. *ACS Appl. Mater. Interfaces* **2012**, *4*, 6273–6279. [CrossRef]

108. Yang, K.; Huang, X.; Xie, L.; Wu, C.; Jiang, P.; Tanaka, T. Core-Shell Structured Polystyrene/BaTiO$_3$ Hybrid Nanodielectrics Prepared by In Situ RAFT Polymerization: A Route to High Dielectric Constant and Low Loss Materials with Weak Frequency Dependence. *Macromol. Rapid Commun.* **2012**, *33*, 1921–1926. [CrossRef]
109. Wu, W.; Liu, X.; Qiang, Z.; Yang, J.; Liu, Y.; Huai, K.; Zhang, B.; Jin, S.; Xia, Y.; Fu, K.K.; et al. Inserting insulating barriers into conductive particle channels: A new paradigm for fabricating polymer composites with high dielectric permittivity and low dielectric loss. *Compos. Sci. Technol.* **2021**, *216*, 109070. [CrossRef]
110. Abutalib, M.M.; Rajeh, A. Boosting optical and electrical characteristics of polyvinyl alcohol/carboxymethyl cellulose nanocomposites by GNPs/MWCNTs fillers as an application in energy storage devices. *Int. J. Energy Res.* **2021**, *46*, 6216–6224. [CrossRef]

Disclaimer/Publisher's Note: The statements, opinions and data contained in all publications are solely those of the individual author(s) and contributor(s) and not of MDPI and/or the editor(s). MDPI and/or the editor(s) disclaim responsibility for any injury to people or property resulting from any ideas, methods, instructions or products referred to in the content.

Article

Synthesis of Monodisperse Mesoporous Carbon Spheres/EPDM Rubber Composites and Their Enhancement Mechanical Properties

Tong Zheng [1], Wenjing Jia [1], Hongjie Meng [2,*], Jiajie Li [1,*] and Xundao Liu [1,*]

[1] School of Material Science and Engineering, University of Jinan, Jinan 250022, China; zhengtong974@163.com (T.Z.)

[2] Shanghai Key Lab of Electrical Insulation and Thermal Aging, School of Chemistry and Chemical Engineering, Center of Hydrogen Science, Shanghai Jiao Tong University, Shanghai 200240, China

* Correspondence: menghongjie@sjtu.edu.cn (H.M.); mse_lijj@ujn.edu.cn (J.L.); mse_liuxundao@ujn.edu.cn (X.L.)

Abstract: Monodisperse mesoporous carbon spheres (MCS) were synthesized and their potential applications in ethylene propylene diene monomer (EPDM) foam were evaluated. The obtained MCS exhibited a high specific surface area ranging from 621-to 735 m^2/g along with large pore sizes. It was observed that the incorporation of MCS into EPDM foam rubber significantly enhances its mechanical properties. The prepared MCS-40 rubber composites exhibit the highest tear strength of 210 N/m and tensile strength of 132.72 kPa, surpassing those of other samples. The enhancement mechanism was further investigated by employing computer simulation technology. The pores within the MCS allowed for the infiltration of EPDM molecular chains, thereby strengthening the interaction forces between the filler and matrix. Moreover, a higher specific surface area resulted in greater adsorption of molecular chains onto the surface of these carbon spheres. This research offers novel insights for understanding the enhancement mechanism of monodisperse mesoporous particles/polymer composites (MCS/EPDM) and highlights their potential application in high-performance rubber composites.

Keywords: monodisperse mesoporous carbon spheres; ethylene propylene diene monomer; composites; mechanical properties; computer simulation

Citation: Zheng, T.; Jia, W.; Meng, H.; Li, J.; Liu, X. Synthesis of Monodisperse Mesoporous Carbon Spheres/EPDM Rubber Composites and Their Enhancement Mechanical Properties. *Polymers* **2024**, *16*, 355. https://doi.org/10.3390/polym16030355

Academic Editors: Yuwei Chen and Yumin Xia

Received: 19 December 2023
Revised: 25 January 2024
Accepted: 25 January 2024
Published: 28 January 2024

Copyright: © 2024 by the authors. Licensee MDPI, Basel, Switzerland. This article is an open access article distributed under the terms and conditions of the Creative Commons Attribution (CC BY) license (https://creativecommons.org/licenses/by/4.0/).

1. Introduction

Rubber materials are extensively employed in both industrial and everyday applications. Ethylene propylene diene monomer (EPDM) rubber, a prevalent type of synthetic rubber, exhibits excellent properties that rely on its internal structure and the additives used, making it increasingly utilized in various technical and industrial applications [1]. The selection of fillers generally impacts both the physical properties and the production costs. Commonly used fillers include carbon black [2], silica [3], calcium carbonate [4], and kaolin [5]. The use of innovative fillers to achieve high-performance rubber offers extensive prospects. The behavior of fillers in elastomer matrices is heavily influenced by the filler's structure (specific surface area, shape, functional groups), the characteristics of the polymer medium (polarity, structure), and the crosslinking system. Recent studies have increasingly focused on mesoporous materials due to their distinctive attributes such as high surface area, large pore volume, and controllable structure [6]. Mesoporous carbon materials, a specific of carbon materials characterized by pore sizes ranging from 2 to 50 nm, are typically synthesized using precursors like rice husk [7], asphalt [8], and phenol-formaldehyde resin [9], followed by physical or chemical activation. Compared to other porous carbon materials such as activated carbon and molecular sieves, mesoporous carbon has gained significant attention in various fields including catalysis, adsorption, gas sensing, and energy conversion and storage due to its exceptional thermal stability, chemical inertness, and excellent conductivity. However, the potential of mesoporous carbon materials in the rubber sector, particularly in EPDM foam materials, remains largely unexplored. The discovery of fullerenes and carbon nanotubes has sparked considerable interest in the exploration of various shaped carbon materials, such as fibers [10], spheres [11], horns [12], and flasks [13]. Extensively studied for decades, spherical carbon materials have found wide-ranging applications in chromatography [14], as catalyst supports, and in rubber reinforcement [15]. Kseniia et al. have reported that the geometric shape of carbon fillers significantly influences elastomer properties [16]. The enhancement effect of carbon materials is influenced by their shape and interaction, which determines their interaction and distribution within the matrix. Carbon materials with different shapes, such as nanodiamonds [17], graphene nanoplates [18], and multi-walled carbon nanotubes [19], exhibit distinct dispersion and orientation states in elastomer composites due to their unique geometric features, thereby affecting mechanical properties including strength, toughness, and stiffness. For example, layered graphene nanoplates can form more effective mechanical reinforcement networks within materials, while nanotubes may provide superior bending strength and toughness. Spherical carbon materials [15] promote better dispersion.

The primary approach for synthesizing mesoporous carbon materials is template casting [20], which involves replicating the internal skeletal structure of various mesoporous inorganic templates in carbon materials. To date, different structures of mesoporous carbon nanospheres have been successfully synthesized using methods such as soft templating [21] and hard templating [22]. For example, Tang et al. reported a soft templating method that involved blending diblock copolymer PEO-b-PS with polydopamine to prepare mesoporous carbon nanospheres [23]. Recently, Wang et al. synthesized mesoporous carbon nanospheres by using polyaniline as a nitrogen source and colloidal silica as a hard template [24]. However, this nanocasting strategy is laborious, intricate, and time-consuming. In recent years, hydrothermal carbonization (HTC) of carbohydrates has gained recognition as an environmentally friendly and sustainable process for synthesizing spherical carbon-based materials [25]. The main objective in preparing carbon sphere materials lies in ensuring their monodispersity while being able to control their outer dimensions and chemical properties [11]. Nevertheless, during practical synthesis, hydrothermally synthesized carbon spheres from sucrose as the carbon source often exhibit interconnections and irregularities. Therefore, achieving simple preparation of monodisperse mesoporous carbon spheres holds significant importance.

Rubber composite materials consist of polymers with molecular weights ranging from several hundred to several thousand and fillers with micro- and nano-scale diameters. A range of EPDM rubber composites containing carbon nanoparticles have been synthesized. Eyssa et al. [26] fabricated sponge ethylene propylene diene rubber (EPDM) nanocomposites based on functionalized multi-walled carbon nanotubes (f-MWCNTs) and foaming agent azodicarbonamide (AZD), which show significantly improved mechanical properties. Shojaei Dindarloo et al. [27] report the effect of various nano-particle types and concentrations on vulcanization and mechanical characteristics of EPDM rubber foam, which demonstrates that the foam's properties were efficiently influenced by both the shapes and content of the nanoparticles in the matrix. Utilizing appropriate nanoparticles is highly advantageous for enhancing the mechanical properties of foamed rubber. Monodispersed mesoporous carbon spheres, due to their uniform size distribution and mesoporous structure, offer significant advantages over other nanoparticles as fillers in EPDM rubber foams. Their unique structural characteristics may lead to more effective stress distribution and enhanced mechanical properties, thereby presenting a novel and potentially more efficient method for improving the performance of EPDM rubber foams. However, due to the significant size difference between the fillers and polymer chains, accurate calculation of the mechanical properties using full atomistic molecular dynamics simulations is not feasible [28]. To overcome this limitation, coarse-grained molecular dynamics simulations were conducted utilizing the Kremer–Grest chain (bead-spring model). In such simulations, each bead in a chain represents multiple monomer units within a molecular model [29]. Although the chemical effects resulting from chain linking are not considered in coarse-grained molecular dynamics simulations, appropriate potential energies between beads (such as bond and nonbond potentials) enable the simulation of chain dynamics at the micro-scale level.

The present study focuses on the synthesis of monodisperse mesoporous carbon spheres and explores their potential application in EPDM rubber foam. To achieve this, uniform-sized and mesoporous carbon spheres were successfully synthesized using a polymer dispersant method. These carbon spheres exhibited high specific surface areas (621–735 m^2/g) and large pore sizes, with adjustable structural features by controlling the amount of sodium polyacrylate. The impact of incorporating these mesoporous carbon spheres as fillers into EPDM foam rubber on mechanical properties was examined. Due to the unique properties of the mesoporous carbon spheres, better integration with the rubber matrix was achieved, resulting in enhanced strength and hardness of the EPDM foam rubber. Furthermore, a coarse-grained molecular dynamics model was established to simulate the stress–strain curve of rubber composite materials under tension, providing insights into the mechanism behind this enhancement effect. This research offers valuable insights for understanding the enhancement mechanism of monodisperse mesoporous particles/polymer composites and highlights their potential application in high-performance rubber composites.

2. Materials and Methods

2.1. Chemicals

Ludox HS-40 (40 wt%, 12 nm), sucrose ($C_{12}H_{22}O_{11}$, 99.5%), and poly (acrylic acid sodium salt) were provided by Sigma Aldrich (St. Louis, MO, USA). EPDM was obtained from the Jilin Chemical Industrial Limited Company of China (Jilin, China). Carbon black N550 was obtained from Cabot Corporation (St. Boston, MA, USA) in the United States; foaming agent AC was acquired from Changzhou Yongxin Fine Chemical Co., Ltd. (Changzhou, China). The antioxidant 2246, zinc oxide (ZnO), stearic acid (SA), accelerator N-cyclo-hexylbenzothiazole-2-sulphenamide (CZ), sulfuric acid (H_2SO_4, 98%), sodium hydroxide (NaOH), and insoluble sulfur (S) are all industrial grade products from Kemiou Chemical Co., Ltd. (Tianjin, China).

2.2. Synthesis of Monodisperse Mesoporous Carbon Sphere

Sucrose was the precursor utilized in this study. Commercial silica sol (Ludox HS-40) was used as the templating agent, maintaining a sucrose/silica dioxide molar ratio of 5.5:1. An amount of 50 mL of 0.66 M sucrose solution requires the addition of 7 mL of Ludox HS-40, concentrated 1.25 mL sulfuric acid was introduced as a catalyst. Continuously stir the mixture for 30 min to ensure the formation of a homogeneous solution. Subsequently, the solution was then pre-carbonized by heating at 100 °C for 24 h. The resulting sludge-like pre-carbonized product was separated and washed with deionized water until neutral pH was achieved. The sludge-like resultant mixture was dried at 100 °C for 6 h and a dark brown powder was obtained. Following that, the product was calcined at 900 °C for 3 h under an inert atmosphere (N_2). Finally, the silica template was removed by NaOH solution (2M) at room temperature, with stir overnight. The mesoporous carbon spheres (MCS) were then washed with ethanol and deionized water.

To prepare monodisperse mesoporous carbon spheres, sodium polyacrylate (0.04% by mass) was added prior to adding the templating agent, followed by stirring for 8 h to ensure complete dissolution before rapidly introducing the templating agent and catalyst into the mixture. The remaining steps were identical to those for the synthesis of MCS. The resulting product is referred to as MCS-X (where X represents the amount of sodium polyacrylate added).

2.3. Synthesis of MCS/EPDM Rubber Composites

MCS/EPDM rubber composites were synthesized via two-roll open mill. First, EPDM (100 phr) was masticated for 2 min using an open two-roll mill (ZG-200L, Dongguan Zhenggong Mechanical and Electrical Equipment Technology Co., Ltd., Dongguan, China), Then, zinc oxide (5 phr), stearic acid (1 phr), carbon black N550 (8 phr), MCS (1phr), antioxidant 2246 (2 phr), accelerator CZ (1 phr), foaming agent AC (4phr), and sulfur (1.6 phr) were sequentially added to the masticated EPDM over a period of 30 min until a homogeneous mixture was formed [30]. Finally, the material was cooled to room temperature and then subjected to foaming in an oven at a heating rate of 0.5 °C/min up to 160 °C.

2.4. Characterization and Performance Tests

2.4.1. Characterization

The textural properties of MCS were evaluated through nitrogen adsorption–desorption isotherms at −196 °C, using a Micromeritics ASAP-2460 adsorption apparatus(Micromeritics Co., Ltd., St. Norcross, GA, USA). Both specific surface area and pore volume were determined using the BET (Brunauer–Emmett–Teller) equation and the single-point method, respectively. Pore size distribution (PSD) curves were evaluated by BJH (Barrett–Joyner–Halenda) method. Furthermore, the t-plot method was employed to determine the micropore volume and mesoporous surface. Scanning electron microscopy (SEM) images were captured by an electron microscope (Hitachi Ltd., Tokyo, Japan, SU8010) at an electron beam voltage of 1.0 kV. The particle size statistics and measurements on the SEM images of the MCS were analyzed using Nano Measurer 1.2.5.

2.4.2. Testing of the MCS/EPDM Rubber Composites

The mechanical properties of MCS/EPDM rubber composites were determined in accordance with the national testing standards of China (GB/T 6344-2008 [31], GB/T 10808-2006 [32], and GB/T 6669-2008 [33]), using a Universal Materials Testing Machine (model CMT6102, Shenzhen Sansi Experimental Equipment Co., Ltd., Shenzhen, City).

Tensile strength and elongation at break are crucial performance parameters for foamed rubber products. The MCS/EPDM rubber composites were made into 13 mm × 152 mm dumbbell-shaped samples with a gauge length of 50 mm. Sample thickness was measured by dial calipers. The universal material testing machine employed in this study operated at a speed of 500 mm/min. A total of ten samples were tested under each loading condition.

Tear strength measures the material's resistance to tearing and it is related to the durability of the foam. The MCS/EPDM rubber composites were made into 25 mm × 25 mm × 125 mm rectangular samples with a 50 mm center cut along. The two resulting strips from the razor notch are pulled apart using the universal material testing machine (Shenzhen Sansi Experimental Equipment Co., Ltd., Shenzhen, China) at a rate of 10 mm/min until a minimum of 20 mm length is torn. The maximum force is then recorded and divided by the sample thickness. Five samples were tested under each loading condition.

The compression strength test measures the load-bearing capacity of the foam after specified conditions of time and temperature. The MCS/EPDM rubber composites were made into 50 mm × 25 mm (diameter × height) samples. The compression ratio was 50%. The speed of the universal material testing machine was 2 mm/min. Three samples were tested under each loading condition.

2.4.3. Coarse-Grained Molecular Dynamics Simulation

The elongation behavior of MCS/EPDM rubber composites was simulated by using the simple bead–spring model. Nonbonded interactions are governed by a Lennard–Jones potential energy function of Formulation (1) [34].

$$U_{LJ(r)} = \begin{cases} 4\varepsilon\left[\left(\frac{\sigma}{r}\right)^{12} - \left(\frac{\sigma}{r}\right)^{6}\right], & r \leq r_{in} \\ \sum_{j=0}^{4} C_j(r - r_{in})^j, & r_{in} < r \leq r_c \\ 0, & r > r_c \end{cases} \quad (1)$$

where ε and σ are the Lennard–Jones parameters for energy and length, respectively, is the distance between any two beads, r_c is a cutoff distance, r_{in} is the inner cutoff distance for a smoothed potential, and the C values are constants calculated so that the force and its first derivative go smoothly from the Lennard–Jones potential at the inner cutoff to zero at the outer cutoff distance.

The value of ε is unity for polymer–polymer and particle–particle interactions, thereby promoting mixing or dispersion of the rods within the polymer matrix. The bonded energy between two consecutive beads in the same chain/particle is given by a harmonic function of the form.

$$U_{bond}(r) = k_r(r - r_0)^2 \quad (2)$$

The rodlike character of the inclusions is enforced through a bending potential of the form.

$$U_{bend}(r) = k_\theta(\theta - \theta_0)^2 \quad (3)$$

In this simulation calculations, where the spring constant is set to $k_r = 10^3$ and $10^4 \times \varepsilon/\sigma^2$ for polymer bonds and nanorod bonds, k_θ is set to 200 ε rad^{-2}. In the coarse-grained model, the box size was set to 50^3, containing 20 MCS units, and a volume fraction of 0.25%. This model was used to simulate and analyze the impact of the filler's specific surface area on the mechanical properties of rubber composites.

3. Results

3.1. Characterization of Mesoporous Carbon Spheres

During the synthesis process, sodium polyacrylate (PAANa) plays a pivotal role in regulating the uniformity of the synthesized mesoporous carbon spheres. Gong et al. [35] have proposed that incorporating a dispersant in the hydrothermal synthesis of carbon spheres facilitates the formation of uniformly sized carbon spheres. To investigate the influence of the dispersants on the synthesis of mesoporous carbon spheres, Ludox HS-40 was selected as the template. After the addition of sodium polyacrylate during the synthesis process, the resulting product was designated as MCS-X, where X represents the quantity of sodium polyacrylate added. This study aims to elucidate how sodium polyacrylate impacts and enhances the uniformity of mesoporous carbon spheres.

Figure 1 illustrates the SEM images of MCS and MCS-40. As shown in Figure 1a, the mesoporous carbon spheres synthesized without the dispersant resulted in a spherical morphology. However, there was noticeable heterogeneity in particle size, accompanied by some degree of adhesion and aggregation. This phenomenon can be attributed to the high temperatures during synthesis, which induced nucleation and subsequent condensation of sucrose molecules, ultimately leading to irregular and crosslinked carbon spheres [36]. In contrast, as shown in Figure 1b, the addition of PAAN facilitated the formation of monodisperse and uniform carbon spheres [37]. The prepared spheres exhibited complete monodispersity with a consistent size (0.9–1.0 μm), and no other shapes of particles or fragments were observed within the samples. Particle size distribution analysis of both types of mesoporous carbon spheres is performed. Figure 2a demonstrates that MCS exhibits a broad range of particle sizes, varying from 0.9 to 4.0 μm, indicating an uneven size distribution within the material. In contrast, Figure 2b illustrates that MCS-40 displays a narrow particle size distribution, centered around 0.9–1.0 µm, suggesting a high level of uniformity in terms of particle size. This narrow distribution signifies the successful synthesis of monodisperse mesoporous carbon spheres.

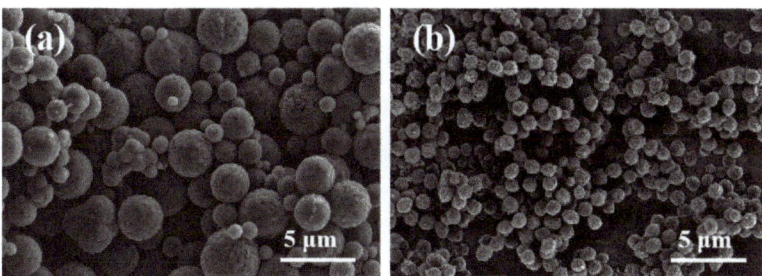

Figure 1. SEM images of mesoporous carbon spheres MCS (**a**) and MCS-40 (**b**).

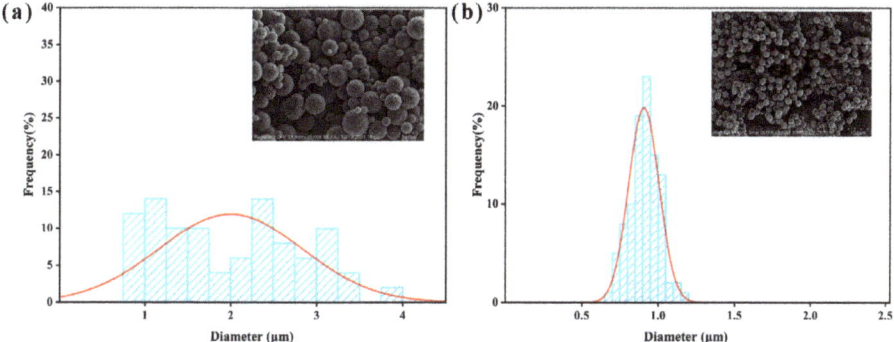

Figure 2. Particle size distribution histograms of (**a**) MCS and (**b**) MCS-40.

To investigate the impact of sodium polyacrylate (PAANa) on the morphology of synthesized mesoporous carbon spheres, varying amounts of PAANa were utilized during the synthesis process. Figure 3 demonstrates a significant enhancement in the uniformity of mesoporous carbon spheres upon the addition of this dispersant, resulting in an overall more uniform particle size. However, with increasing amounts of PAANa, agglomeration among the mesoporous carbon spheres was observed. Specifically, when 40 mg of dispersant was added, some carbon spheres showed slight fragmentation, which affected their spherical morphology. This phenomenon can be attributed to excessive aggregation of excessive dispersant that impedes proper encapsulation and carbonization processes for the carbon source, ultimately leading to surface damage in the formed mesoporous carbon spheres [35].

3.2. Morphology of the Foams

The SEM images of the foam samples are shown in Figure 4 which revealed the presence of both open cells and closed cells within the porous structure of the foams. The white parts represent closed pores and the dark part represents open pores. The semi-open cell structure represented the edges and surfaces along with the existence of some holes in the cell walls. These pores allow gas molecules to pass through the continuous phase, thereby creating fluidity through the foam to some extent [35]. Furthermore, nano-fillers acted as active nucleating agents, providing more bubble sites. The effect of MCS on the pore structure of EPDM foams was not significant. The effect of MCS on the mechanical properties of EPDM foams is mainly due to the interaction between MCS and rubber matrix.

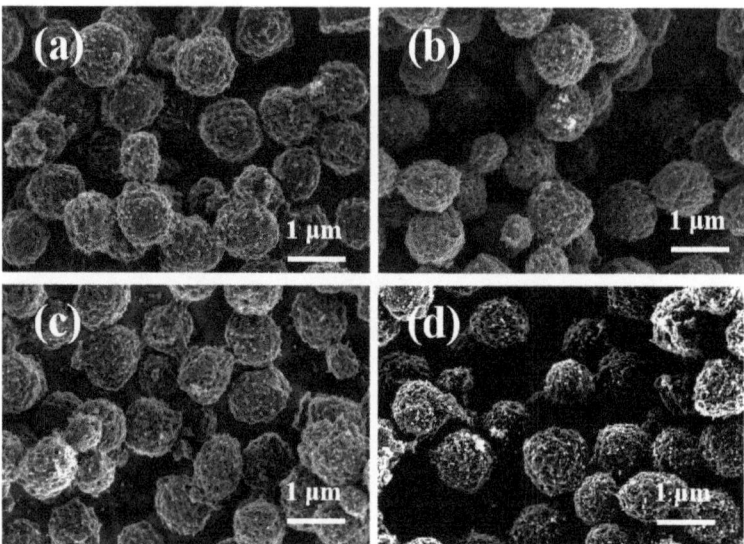

Figure 3. SEM images of mesoporous carbon spheres MCS-X (X is the amount of sodium polyacrylate added in mg). (**a**) MCS-20; (**b**) MCS-25; (**c**) MCS-30; and (**d**) MCS-40.

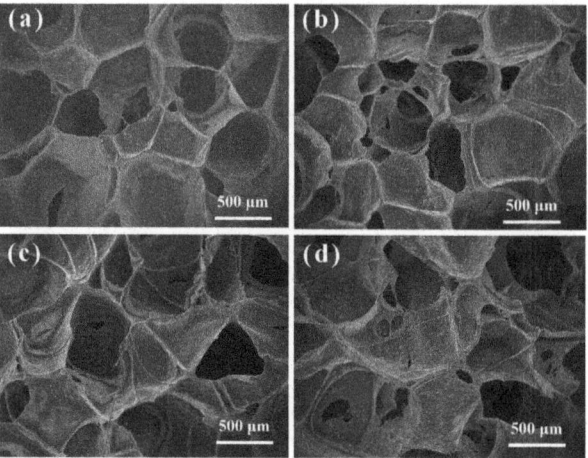

Figure 4. SEM images of the MCS/EPDM nanocomposite foam samples containing (**a**) MCS-20, (**b**) MCS-25, (**c**) MCS-30, and (**d**) MCS-40.

3.3. Specific Surface Area and Pore Size Analysis of Mesoporous Carbon Spheres

Table 1 presents the characterization results of mesoporous carbon spheres (MCS) synthesized with varying weights of sodium polyacrylate (PAANa). Figure 5 shows the nitrogen adsorption–desorption isotherms and the corresponding pore size distribution curves. As depicted in Figure 5a, MCS-X exhibits Type IV BET adsorption isotherms according to IUPAC classification, accompanied by H-2 hysteresis loops, which are typical characteristics of mesoporous materials. Furthermore, the hysteresis loop at high pressure ($P/P0 = 0.89$–0.98) reflects the existence of large mesopores, which could be attributed to the inter-particle packing between the mesoporous carbon nanospheres [38]. Notably, as shown in Figure 5a, both the overall specific surface area and the low-pressure range micropore area increase with the increasing dosage of dispersant PAANa. The microporous specific surface area was determined using the t-plot method, which involves calculating the microporous specific surface area by fitting a linear curve to the adsorption isotherm's low-pressure region in the microporous adsorption region. For calculating the external surface area (St-plot) from the slope of the linear fit, the relation proposed by Harkins and Jura [39] was employed as the standard reference t-curve, combined with the data presented in Table 1. By analyzing the textural properties of mesoporous carbon spheres with added PAANa, it is evident that the Brunauer–Emmett–Teller (BET)-specific surface areas of the mesoporous carbon spheres are, respectively, 621 m^2/g, 673 m^2/g, 695 m^2/g, and 735 m^2/g, demonstrating a consistent upward trend. The mesopore size distribution curves derived from the Barrett—Joyner–Hallenda (BJH) method [40] show that the MCS-40 mesopore diameters are 3 nm and 8 nm, respectively. The micropore-specific surface areas are measured as 631 m^2/g, 582 m^2/g, 563 m^2/g, and 529 m^2/g. Furthermore, the pore volume exhibits an increase upon the addition of sodium polyacrylate.

Table 1. Textural properties of mesoporous carbon spheres added with PAANa.

	S_{BET} [a] (m^2/g)	$S_{t\text{-plot}}$ [b] (m^2/g)	V_p [c] (cm^3/g)	Pore Size (nm)
MCS-20	621	631	0.33	3.0; 7.0
MCS-25	673	582	0.25	3.0; 9.0
MCS-30	695	563	0.40	3.0; 8.0
MCS-40	735	529	0.52	3.0; 8.0

[a] Specific surface area was calculated by BET method based on nitrogen adsorption isotherm. [b] Specific surface area of micropores calculated by the t-plot method. [c] Total pore volume calculated from nitrogen adsorption data with P/P0 = 0.99.

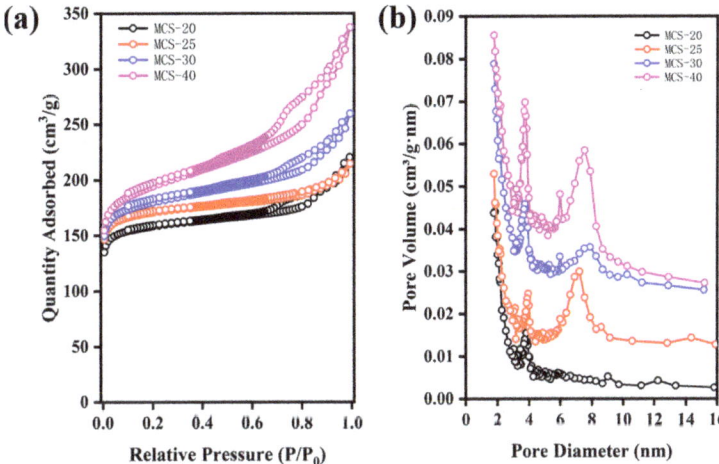

Figure 5. N$_2$ adsorption and desorption curves of MCS with different contents of sodium polyacrylate: (**a**) N$_2$ adsorption and desorption curves and (**b**) pore size distribution.

3.4. Mechanical Properties

The impact of different types of mesoporous carbon spheres (MCS) on the mechanical properties of MCS/EPDM rubber composites is shown in Figure 6, while specific values are provided in Table 2. The mechanical properties of EPDM foam materials are improved with the addition of MCS. Figure 6a demonstrates the elongation at break and the tensile strength of EPDM foam materials upon the addition of MCS-X. As shown in Figure 6a, MCS-40 exhibits superior tensile strength, albeit with slightly compromised elongation at break performance. The favorable compatibility between the carbon material and the polymer matrix facilitates effective stress transfer, thereby enhancing the mechanical properties of the composite.

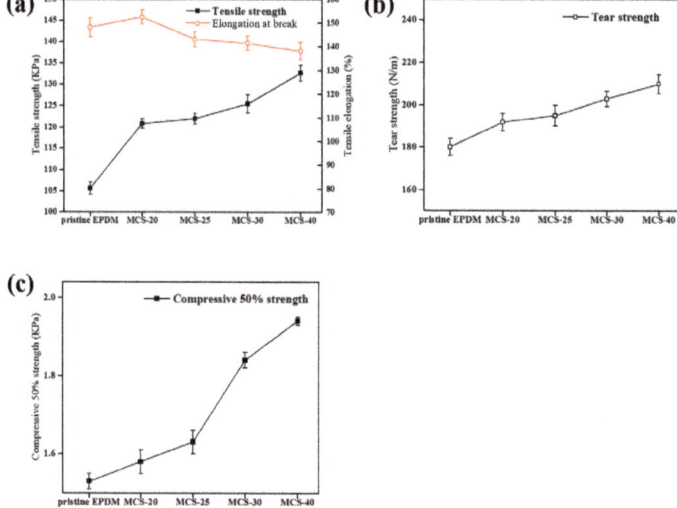

Figure 6. The effect of MCS-X on the mechanical properties of MCS/EPDM rubber composites: (**a**) tensile strength and elongation at break. Values are means ± s.e.m.; n = 10; (**b**) tear strength. Values are means ± s.e.m.; n = 5; (**c**) compress 50% strength. Values are means ± s.e.m.; n = 3.

Table 2. Mechanical properties of MCS/EPDM rubber composites.

	Filler Type				
	Pristine EPDM	**MCS-20**	**MCS-25**	**MCS-30**	**MCS-40**
Tensile strength (KPa)	105.66 (±1.42)	120.88 (±1.17)	122.00 (±1.24)	125.45 (±2.15)	132.72 (±1.83)
Elongation at break (%)	148.00 (±4.07)	152.43 (±2.99)	143.00 (±3.08)	141.45 (±2.97)	138.15 (±3.62)
Tear strength (N/m)	180 (±4.12)	192 (±4.17)	195 (±4.76)	203 (±3.63)	210 (±4.5)
Compressive 50% strength (KPa)	1.53 (±0.02)	1.58 (±0.03)	1.63 (±0.03)	1.84 (±0.02)	1.94 (±0.01)

In Figure 6b, it can be observed that MCS-40 possesses the highest tear strength due to the effective restriction of rubber chain segmental motion within the mesoporous channels [41]. Furthermore, as depicted in Figure 6c, MCS-40 displays the best compression strength, which is attributed to the large specific surface area of MCS-40, to allow for better bonding with the rubber matrix. The addition of PAANa results in a larger specific surface area of MCS, which promotes increased binding glue formation leading to enhanced reinforcement, elevated tensile strength, and greater load-bearing capacity.

In the coarse-grained model, the forces utilized include the harmonic potential and the Lennard–Jones potential. The harmonic potential is employed to simulate bonded interactions, while the Lennard–Jones potential represents non-bonded interactions. These forces play a crucial role in simulating the behavior and properties of filled cross-linked rubber [35]. Moreover, within this coarse-grained model, variations in filler-specific surface area directly impact the non-bonded interactions, whereby larger surface areas correspond to stronger forces [42].

Figure 7 shows the stress–strain curves in the uniaxial elongation simulation of the filled crosslinked polymer model. During the elongation process, the stress was influenced by the strength of the filler–polymer interaction, with a stronger interaction leading to higher stress. The strongly attractive filler–polymer interaction caused a shift in the stress upturn point towards lower elongation values. Moreover, materials with larger specific surface areas provided more binding sites, thereby enhancing the interaction forces between the filler and rubber [43]. As shown in Figure 7, MCS-40 with a larger specific surface area exhibits greater stress and a lower elongation rate compared to other samples under identical conditions.

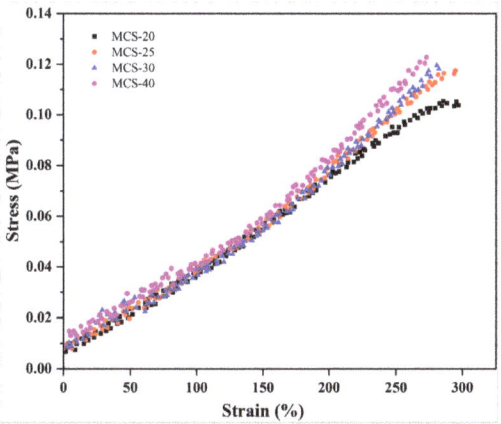

Figure 7. Simulated stress–strain elongation curves in the uniaxial of MCS/EPDM rubber composites.

Figure 8 shows the images of the coarse-grained model at 0% and 200% stretch, where the green balls represent the synthetic monodisperse mesoporous carbon spheres. Surrounding these carbon spheres are white polymer chains of EPDM, simulating the adsorption of the EPDM molecular chains onto the surface of carbon spheres. The pore structure on the carbon sphere's surface facilitates enhanced interaction forces by allowing penetration of EPDM molecular chains into these pores. Consequently, a larger specific surface area leads to the increased adsorption of molecular chains onto the surface of this carbon ball. Adding the filler resulted in a non-linear viscoelastic behavior, known as the Payne effect. The higher the surface area the higher the Payne effect, which corresponds to the extent of filler networking. The influence of the mesoporous carbon sphere surface area on the formation of a filler network can be explained as follows: At fixed levels of structure and filler loading, both the aggregate size and the inter-aggregate distance decrease with, respectively, increasing surface area [44]. The smaller the inter-aggregate distance the higher the probability for the formation of a filler network. Consequently, when the surface area increases, the extent of the packing network is more pronounced, which is manifested as an increase in tensile strength. During the stretching process, the surface of the molecular chain is also stretched; thus, stronger surface forces result in higher tensile strength while weaker ones lead to lower tensile strength. At 200% stretching, both the orientation of molecular chains and motion-induced changes in the mesoporous carbon sphere's molecular weight align with the direction of applied tensile force.

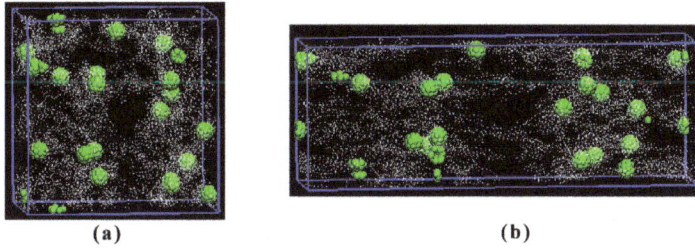

Figure 8. Coarse-grained models of the composites with MCS-40, (**a**) 0%, and (**b**) 200% elongation.

4. Conclusions

In summary, new monodisperse mesoporous particles/polymer composites (MCS/EPDM) were successfully synthesized by incorporating monodisperse mesoporous carbon spheres (MCS) into ethylene propylene diene monomer (EPDM) foam. The synthesized MCS has a uniform size and mesoporous structure, as well as a high specific surface area. The infiltration of EPDM molecular chains within the pores of MCS strengthened the interaction forces between the filler and matrix. The enhancement mechanism was further investigated by employing computer simulation technology. The simulation of the tensile situation based on coarse-grained molecular dynamics revealed that a higher specific surface area facilitates enhanced accessibility of polymer molecular into the pores of the mesoporous carbon spheres, thus leading to a stronger interaction force between the mesoporous carbon spheres and the polymer chains, ultimately resulting in an enhancement in the tensile strength. This research provides new insights into understanding the enhancement mechanism of MCS/EPDM composites and highlights their potential application in high-performance rubber composites.

Author Contributions: Conceptualization, validation, formal analysis, and methodology, T.Z.; investigation, resources, and data curation, H.M.; writing—original draft preparation, W.J.; writing—review and editing, J.L. investigation and writing—review and editing. X.L. All authors have read and agreed to the published version of the manuscript.

Funding: This work was supported by the Science and Technology Program of the University of Jinan (No.: XKY2105, XKY2103).

Data Availability Statement: The data that support the findings of this study are available from the first author, Tong Zheng, upon reasonable request (due to privacy).

Acknowledgments: The authors acknowledge Yue Li for his support in constructing the coarse-grained model.

Conflicts of Interest: The authors declare no conflicts of interest.

References

1. Gatos, K.G.; Karger-Kocsis, J. Effects of Primary and Quaternary Amine Intercalants on the Organoclay Dispersion in a Sulfur-Cured EPDM Rubber. *Polymer* **2005**, *46*, 3069–3076. [CrossRef]
2. Auer, E.; Freund, A.; Pietsch, J.; Tacke, T. Carbons as Supports for Industrial Precious Metal Catalysts. *Appl. Catal. A Gen.* **1998**, *173*, 259–271. [CrossRef]
3. Yang, R.; Song, Y.; Zheng, Q. Payne Effect of Silica-Filled Styrene-Butadiene Rubber. *Polymer* **2017**, *116*, 304–313. [CrossRef]
4. Cai, H.-H.; Li, S.-D.; Tian, G.-R.; Wang, H.-B.; Wang, J.-H. Reinforcement of Natural Rubber Latex Film by Ultrafine Calcium Carbonate. *J. Appl. Polym. Sci.* **2003**, *87*, 982–985. [CrossRef]
5. Sreelekshmi, R.V.; Sudha, J.D.; Menon, A.R.R. Novel Organomodified Kaolin/Silica Hybrid Fillers in Natural Rubber and Its Blend with Polybutadiene Rubber. *Polym. Bull.* **2017**, *74*, 783–801. [CrossRef]
6. Sun, Y.; Ma, K.; Kao, T.; Spoth, K.A.; Sai, H.; Zhang, D.; Kourkoutis, L.F.; Elser, V.; Wiesner, U. Formation Pathways of Mesoporous Silica Nanoparticles with Dodecagonal Tiling. *Nat. Commun.* **2017**, *8*, 252. [CrossRef]
7. Wu, M.; Li, L.; Liu, J.; Li, Y.; Ai, P.; Wu, W.; Zheng, J. Template-Free Preparation of Mesoporous Carbon from Rice Husks for Use in Supercapacitors. *New Carbon Mater.* **2015**, *30*, 471–475. [CrossRef]
8. Zhang, X.; Zhong, C.; Lin, Q.; Luo, S.; Zhang, X.; Fang, C. Facile Fabrication of Graphitic Mesoporous Carbon with Ultrathin Walls from Petroleum Asphalt. *J. Anal. Appl. Pyrolysis* **2017**, *126*, 154–157. [CrossRef]
9. Xia, X.; Cheng, C.-F.; Zhu, Y.; Vogt, B.D. Ultrafast Microwave-Assisted Synthesis of Highly Nitrogen-Doped Ordered Mesoporous Carbon. *Microporous Mesoporous Mater.* **2021**, *310*, 110639. [CrossRef]
10. Saito, N.; Aoki, K.; Usui, Y.; Shimizu, M.; Hara, K.; Narita, N.; Ogihara, N.; Nakamura, K.; Ishigaki, N.; Kato, H. Application of Carbon Fibers to Biomaterials: A New Era of Nano-Level Control of Carbon Fibers after 30-Years of Development. *Chem. Soc. Rev.* **2011**, *40*, 3824–3834. [CrossRef]
11. Deshmukh, A.A.; Mhlanga, S.D.; Coville, N.J. Carbon Spheres. *Mater. Sci. Eng. R Rep.* **2010**, *70*, 1–28. [CrossRef]
12. Du, J.; Liu, Z.; Li, Z.; Han, B.; Sun, Z.; Huang, Y. Carbon Onions Synthesized via Thermal Reduction of Glycerin with Magnesium. *Mater. Chem. Phys.* **2005**, *93*, 178–180. [CrossRef]
13. Wang, Z.L.; Yin, J.S. Graphitic Hollow Carbon Calabashes. *Chem. Phys. Lett.* **1998**, *289*, 189–192. [CrossRef]
14. Wei, Z.; Ren, H.; Wang, S.; Qiu, H.; Liu, X.; Jiang, S. Facile Preparation of Well Dispersed Uniform, Porous Carbon Microspheres and Their Use as a New Chromatographic Adsorbent. *Mater. Lett.* **2013**, *105*, 144–147. [CrossRef]

15. Szadkowski, B.; Marzec, A.; Zaborski, M. Effect of Different Carbon Fillers on the Properties of Nitrile Rubber Composites. *Compos. Interfaces* 2019, *26*, 729–750. [CrossRef]
16. Shakun, A.; Sarlin, E.; Vuorinen, J. Material-Related Losses of Natural Rubber Composites with Surface-Modified Nanodiamonds. *J. Appl. Polym. Sci.* 2020, *137*, 48629. [CrossRef]
17. Mokhireva, K.A.; Svistkov, A.L.; Solod'ko, V.N.; Komar, L.A.; Stöckelhuber, K.W. Experimental Analysis of the Effect of Carbon Nanoparticles with Different Geometry on the Appearance of Anisotropy of Mechanical Properties in Elastomeric Composites. *Polym. Test.* 2017, *59*, 46–54. [CrossRef]
18. Vozniakovskii, A.A.; Vozniakovskii, A.P.; Kidalov, S.V.; Otvalko, J.; Neverovskaia, A.Y. Characteristics and Mechanical Properties of Composites Based on Nitrile Butadiene Rubber Using Graphene Nanoplatelets. *J. Compos. Mater.* 2020, *54*, 3351–3364. [CrossRef]
19. Li, F.; Lu, Y.; Liu, L.; Zhang, L.; Dai, J.; Ma, J. Relations between Carbon Nanotubes' Length and Their Composites' Mechanical and Functional Performance. *Polymer* 2013, *54*, 2158–2165. [CrossRef]
20. Kresge, C.T.; Leonowicz, M.E.; Roth, W.J.; Vartuli, J.C.; Beck, J.S. Ordered Mesoporous Molecular Sieves Synthesized by a Liquid-Crystal Template Mechanism. *Nature* 1992, *359*, 710–712. [CrossRef]
21. Zhao, D.; Feng, J.; Huo, Q.; Melosh, N.; Fredrickson, G.H.; Chmelka, B.F.; Stucky, G.D. Triblock Copolymer Syntheses of Mesoporous Silica with Periodic 50 to 300 Angstrom Pores. *Science* 1998, *279*, 548–552. [CrossRef] [PubMed]
22. Ryoo, R.; Joo, S.H.; Jun, S. Synthesis of Highly Ordered Carbon Molecular Sieves via Template-Mediated Structural Transformation. *J. Phys. Chem. B* 1999, *103*, 7743–7746. [CrossRef]
23. Tang, J.; Liu, J.; Li, C.; Li, Y.; Tade, M.O.; Dai, S.; Yamauchi, Y. Synthesis of Nitrogen-doped Mesoporous Carbon Spheres with Extra-large Pores through Assembly of Diblock Copolymer Micelles. *Angew. Chem. Int. Ed.* 2015, *54*, 588–593. [CrossRef] [PubMed]
24. Wang, G.; Sun, Y.; Li, D.; Liang, H.; Dong, R.; Feng, X.; Müllen, K. Controlled Synthesis of N-doped Carbon Nanospheres with Tailored Mesopores through Self-assembly of Colloidal Silica. *Angew. Chem.* 2015, *127*, 15406–15411. [CrossRef]
25. Wang, Q.; Li, H.; Chen, L.; Huang, X. Monodispersed Hard Carbon Spherules with Uniform Nanopores. *Carbon* 2001, *39*, 2211–2214. [CrossRef]
26. Eyssa, H.M.; Afifi, M.; Moustafa, H. Improvement of the Acoustic and Mechanical Properties of Sponge Ethylene Propylene Diene Rubber/Carbon Nanotube Composites Crosslinked by Subsequent Sulfur and Electron Beam Irradiation. *Polym. Int.* 2023, *72*, 87–98. [CrossRef]
27. Shojaei Dindarloo, A.; Karrabi, M.; Hamid, M.; Ghoreishy, R. Various Nano-Particles Influences on Structure, Viscoelastic, Vulcanization and Mechanical Behaviour of EPDM Nano-Composite Rubber Foam. *Plast. Rubber Compos.* 2019, *48*, 218–225. [CrossRef]
28. Yagyu, H. Simulations of the Effects of Filler Aggregation and Filler-Rubber Bond on the Elongation Behavior of Filled Cross-Linked Rubber by Coarse-Grained Molecular Dynamics. *Soft Mater.* 2017, *15*, 263–271. [CrossRef]
29. Yagyu, H.; Utsumi, T. Coarse-Grained Molecular Dynamics Simulation of Nanofilled Crosslinked Rubber. *Comput. Mater. Sci.* 2009, *46*, 286–292. [CrossRef]
30. Zhong, B.; Zeng, X.; Chen, W.; Luo, Q.; Hu, D.; Jia, Z.; Jia, D. Nonsolvent-Assisted Surface Modification of Silica by Silane and Antioxidant for Rubber Reinforcement. *Polym. Test.* 2019, *78*, 105949. [CrossRef]
31. GB/T 6344-2008; Flexible cellular polymeric materials—Determination of tensile strength and elongation at break. Standardization Administration of the People's Republic of China: Beijing, China, 2008.
32. GB/T 10808-2006; Flexible cellular polymeric materials—Determination of tear strength. Standardization Administration of the People's Republic of China: Beijing, China, 2006.
33. GB/T 6669-2008; Flexible cellular polymeric materials—Determination of compression set. Standardization Administration of the People's Republic of China: Beijing, China, 2008.
34. Toepperwein, G.N.; Karayiannis, N.C.; Riggleman, R.A.; Kröger, M.; De Pablo, J.J. Influence of Nanorod Inclusions on Structure and Primitive Path Network of Polymer Nanocomposites at Equilibrium and Under Deformation. *Macromolecules* 2011, *44*, 1034–1045. [CrossRef]
35. Gong, Y.; Xie, L.; Li, H.; Wang, Y. Sustainable and Scalable Production of Monodisperse and Highly Uniform Colloidal Carbonaceous Spheres Using Sodium Polyacrylate as the Dispersant. *Chem. Commun.* 2014, *50*, 12633–12636. [CrossRef]
36. Titirici, M.-M.; Antonietti, M.; Baccile, N. Hydrothermal Carbon from Biomass: A Comparison of the Local Structure from Poly- to Monosaccharides and Pentoses/Hexoses. *Green Chem.* 2008, *10*, 1204–1212. [CrossRef]
37. Álvarez-Láinez, M.; Rodríguez-Pérez, M.A.; de Saja, J.A. Acoustic Absorption Coefficient of Open-Cell Polyolefin-Based Foams. *Mater. Lett.* 2014, *121*, 26–30. [CrossRef]
38. Fang, Y.; Gu, D.; Zou, Y.; Wu, Z.; Li, F.; Che, R.; Deng, Y.; Tu, B.; Zhao, D. A Low-Concentration Hydrothermal Synthesis of Biocompatible Ordered Mesoporous Carbon Nanospheres with Tunable and Uniform Size. *Angew. Chem. Int. Ed.* 2010, *49*, 7987–7991. [CrossRef]
39. Jura, G.; Harkins, W.D. Surfaces of Solids. XI. Determination of the Decrease (π) of Free Surface Energy of a Solid by an Adsorbed Film. *J. Am. Chem. Soc.* 1944, *66*, 1356–1362. [CrossRef]
40. Xu, F.; Tang, Z.; Huang, S.; Chen, L.; Liang, Y.; Mai, W.; Zhong, H.; Fu, R.; Wu, D. Facile Synthesis of Ultrahigh-Surface-Area Hollow Carbon Nanospheres for Enhanced Adsorption and Energy Storage. *Nat. Commun.* 2015, *6*, 7221. [CrossRef]

41. Ji, X.; Hampsey, J.E.; Hu, Q.; He, J.; Yang, Z.; Lu, Y. Mesoporous Silica-Reinforced Polymer Nanocomposites. *Chem. Mater.* **2003**, *15*, 3656–3662. [CrossRef]
42. Omnès, B.; Thuillier, S.; Pilvin, P.; Gillet, S. Non-Linear Mechanical Behaviour of Carbon Black Reinforced Elastomers: Experiments and Multiscale Modelling. *Plast. Rubber Compos.* **2008**, *37*, 251–257. [CrossRef]
43. Fu, Q.; Yang, Z.; Jia, H.; Wen, Y.; Luo, Y.; Ding, L. Integration of Experimental Methods and Molecular Dynamics Simulations for a Comprehensive Understanding of Enhancement Mechanisms in Graphene Oxide (GO)/Rubber Composites. *J. Polym. Res.* **2023**, *30*, 277. [CrossRef]
44. Fröhlich, J.; Niedermeier, W.; Luginsland, H.-D. The Effect of Filler–Filler and Filler–Elastomer Interaction on Rubber Reinforcement. *Compos. Part A Appl. Sci. Manuf.* **2005**, *36*, 449–460. [CrossRef]

Disclaimer/Publisher's Note: The statements, opinions and data contained in all publications are solely those of the individual author(s) and contributor(s) and not of MDPI and/or the editor(s). MDPI and/or the editor(s) disclaim responsibility for any injury to people or property resulting from any ideas, methods, instructions or products referred to in the content.

Article

Systematic Investigation of the Degradation Properties of Nitrile-Butadiene Rubber/Polyamide Elastomer/Single-Walled Carbon Nanotube Composites in Thermo-Oxidative and Hot Oil Environments

Guangyong Liu [1,†], Huiyu Wang [1,†], Tianli Ren [2], Yuwei Chen [1] and Susu Liu [1,*]

[1] Key Laboratory of Rubber-Plastics of Ministry of Education, Qingdao University of Science & Technology, Qingdao 266042, China; liuguangyong@qust.edu.cn (G.L.); 15666426387@163.com (H.W.); yuweichen@qust.edu.cn (Y.C.)

[2] Mississippi Polymer Institute, University of Southern Mississippi, Hattiesburg, MS 39406, USA; tianli.ren@usm.edu

* Correspondence: lss8130@163.com

† These authors contributed equally to this work.

Citation: Liu, G.; Wang, H.; Ren, T.; Chen, Y.; Liu, S. Systematic Investigation of the Degradation Properties of Nitrile-Butadiene Rubber/Polyamide Elastomer/Single-Walled Carbon Nanotube Composites in Thermo-Oxidative and Hot Oil Environments. *Polymers* **2024**, *16*, 226. https://doi.org/10.3390/polym16020226

Academic Editor: Changwoon Nah

Received: 11 December 2023
Revised: 4 January 2024
Accepted: 10 January 2024
Published: 12 January 2024

Copyright: © 2024 by the authors. Licensee MDPI, Basel, Switzerland. This article is an open access article distributed under the terms and conditions of the Creative Commons Attribution (CC BY) license (https://creativecommons.org/licenses/by/4.0/).

Abstract: The physical blending method was used in order to prepare nitrile-butadiene rubber/polyamide elastomer/single-walled carbon nanotube (NBR/PAE/SWCNT) composites with better thermal-oxidative aging resistance. The interactions between SWCNTs and NBR/PAE were characterized using the Moving Die Rheometer 2000 (MDR 2000), rheological behavior tests, the equilibrium swelling method, and mechanical property tests. The 100% constant tensile stress and hardness of NBR/PAE/SWCNT composites increased from 2.59 MPa to 4.14 MPa and from 62 Shore A to 69 Shore A, respectively, and the elongation decreased from 421% to 355% with increasing SWCNT content. NBR/PAE/SWCNT composites had improved thermal-oxidative aging resistance due to better interactions between SWCNTs and NBR/PAE. During the aging process, the tensile strength and elongation at break decreased with the increase in aging time compared to the unaged samples, and the constant tensile stress gradually increased. There was a more significant difference in the degradation of mechanical properties when aged in a variety of oils. The 100% constant tensile stress of NBR/PAE/SWCNT composites aged in IRM 903 gradually increased with aging time while it gradually decreased in biodiesel. The swelling index gradually increased with increasing SWCNT content. Interestingly, the swelling index of the composites in cyclohexanone decreased with the increase in SWCNT content. The reasons leading to different swelling behaviors when immersed in different kinds of liquids were investigated using the Hansen solubility parameter (HSP) method, which provides an excellent guide for the application of some oil-resistant products.

Keywords: nitrile-butadiene rubber; polyamide elastomer; single-walled carbon nanotubes; composite; swelling; aging

1. Introduction

Nitrile butadiene rubber (NBR) is an organic polymer elastic compound with superior abrasion resistance, processability, and excellent oil resistance [1–4]. Hence, NBR materials are widely used in the manufacture of some seals, fuel hoses, and other oilfield rubber products [5–9]. Generally, in real conditions, NBR seals are frequently used in severe applications, so the degradation of NBR in these environments is unavoidable [10,11]. Most seal failures are caused by rubber aging, which leads to substantial economic losses. Bingqi Jiang et al. [12] investigated the effect of thermal aging in oil on the friction and wear properties of NBR. Boli et al. [13] studied NBR seals exposed to different aging atmospheres and evaluated the mechanical properties, tribological behavior, and energy dissipation of aged NBR. In the study, thermo-oxidative aging significantly degraded the mechanical properties and greatly affected the hysteresis characteristics during friction, further causing degradation.

As rubber products become more widespread in today's society, the requirements for rubber product performance are becoming increasingly stringent. For environmental protection and sustainable development, incorporating some filler into the polymer matrix plays an ever more critical role in reducing costs and improving performance [14–18]. Emad S. Shafik et al. [19] used red brick waste (RBW) powder as a reinforcing filler in NBR to prepare eco-friendly composites. The findings indicated that such composites could be used for insulation and antistatic applications. In addition, magnetic measurements have shown superparamagnetic behavior in NBR/RBW composites. Dawei Tang et al. [20] studied the incorporation of waste brick powder as a filler in SBR to enhance its mechanical properties. This was performed to reduce the polymer product's cost. Xumin Zhang et al. [21] reported that a novel filler, graphene oxide, enhanced the compatibility and mechanical properties of carboxylated nitrile butadiene rubber/styrene butadiene rubber. The results showed that the addition of graphene oxide could significantly improve the thermal and mechanical properties of the blends as well as their compatibility. Shubham C. Ambilkar et al. [22] employed three different surface modifiers, namely sodium dodecyl sulfate (SDS, a surfactant), 3-(trimethoxy silyl) propyl methacrylate (MPS, an organosilane), and tris (hydroxymethyl) aminomethane (Tris, an amine buffer), to modify the surface of sol–gel derived from in situ generated zirconia and investigated the thermal, morphological, mechanical, swelling, rheological, and dielectric properties. This study revealed that Tris could be a potential surface modifier for metal oxides like zirconia, as an alternative to organosilane, to reinforce the elastomer matrices. S. Utrera-Barrios et al. [23] introduced an innovative approach in which waste parts from toner cartridges were valorized to develop (recyclable and) self-healing elastomeric composite materials. The study found that the thermoplastic elastomers formed from high-impact polystyrene and carboxylated nitrile butadiene rubber from toner cartridge waste had high mechanical properties and self-healing ability. Moreover, an increase in toner content (up to 20 phr) resulted in an optimal balance between tensile strength and self-healing capacity. Fanghui Wang et al. [24] prepared carbon nanofibers–silica (CNFs–SiO_2) nanocomposites by the self-assembly method, which were used as filler to obtain CNFs–SiO_2/HNBR composites. These results indicated that CNFs–SiO_2 as a reinforcing filler could significantly improve the rubber mechanical properties. SWCNTs exhibit an unusual one-dimensional tubular structure, high specific area, and excellent mechanical, electrical, thermal, and chemical properties [25–27]. It is because of these properties that SWCNTs are widely used in the fabrication of polymer composites for different applications. Kazufumi Kobashi et al. [28] proposed a method to control the dendritic structure of long carbon nanotubes (CNTs). The results showed that the electrical conductivity of carbon nanotube rubber composites could be improved by controlling the dendritic structure of CNTs. Jabulani I. Gumede et al. [29] investigated the effect of SWCNTs on the vulcanization properties and mechanical properties of recycled rubber (RR)/natural rubber (NR) blends and found that the addition of SWCNTs decreased the minimum torque, increased the scorch time as well as the vulcanization time of the RR/NR blends, and enhanced the hardness of the blends.

The effect of polyamide elastomer (PAE) on the thermal-oxidative aging resistance of NBR/PAE blends has been previously studied [30]. FTIR, DMA, SEM, and equilibrium swelling methods were used to analyze the interaction between NBR and PAE. It was found that NBR/PAE blends exhibited better tensile strength retention at high temperature. The role of PAE in improving the high-temperature aging resistance of NBR/PAE blends was also explored. Finally, the swelling behavior of rubbers in biodiesel and IRM 903 was measured. Based on earlier studies, it was found that the addition of SWCNTs resulted in the improvement of tensile strength, modulus, hardness, and electrical conductivity of the blends, while the elongation at break was slightly reduced [31–34]. Previous studies have been conducted on the effect of SWCNTs on the mechanical and electrical conductivity properties of rubber. However, no one has investigated the changes in the properties of composites containing SWCNTs during air and oil aging.

In this study, NBR/PAE/SWCNT composites with different SWCNT contents were prepared by the mechanical blending method. The interactions between SWCNTs and NBR/PAE were investigated by its vulcanization properties, using a Rubber Processing Analyzer (RPA2000), the equilibrium swelling method, and mechanical property tests. Since NBR composites are mainly used in seals, high temperatures and oil liquids can dramatically affect their properties. The degradation of mechanical properties was evaluated by performing tensile and other performance tests after thermo-oxidative aging and thermal oil aging. Moreover, the HSP method was used to explore the reasons for the changes in the swelling index of the NBR/PAE/SWCNT composites after immersion in different types of liquids, which provides an excellent guide for the study of some oil-resistant products.

2. Experiment

2.1. Materials and Preparation of NBR/PAE/SWCNT Composites

NBR-3445 (M_W = 100,000 g/mol) was purchased from Arlanxeo (The Hague, The Netherlands), with 34% acrylonitrile content. Polyamide elastomer (M_W = 40,000 g/mol) was supplied by Arkema France (Paris Villepinte, France), with a melting point of 170–180 °C. The curing agents and chemicals, such as sulfur, zinc oxide (ZnO), stearic acid, tetramethylthiuram disulfide (TMTD), 2,2′-dithiobis(benzothiazole)(DM), antioxidant poly (1,2-dihydro-2,2,4-trimethylquinoline) (TMQ), and N-1,3-dimethylbutyl-N'-phenyl-p-phenylenediamine (DMPPD) were supplied by Rhein Chemie, (Qingdao, China). Carbon black (N330) was produced by Cabot (Alpharetta, GA, USA). The SWCNTs were produced by OCSiAl (Novosibirsk, Russia). The used SWCNTs had a diameter of 2 nm, length of 5 μm, and purity of 95%. Table 1 shows the specimen formulas. The structural formulas of NBR and PAE are shown in Figure 1.

Table 1. Recipe for the NBR/PAE/SWCNT composites.

Sample/phr	1	2	3	4
NBR	100	100	100	100
PAE	20	20	20	20
N330	30	30	30	30
TMTD	0.5	0.5	0.5	0.5
DM	2	2	2	2
DMPPD	1	1	1	1
TMQ	0.5	0.5	0.5	0.5
ZnO	5	5	5	5
Stearic acid	1	1	1	1
Sulfur	1.5	1.5	1.5	1.5
SWCNT	0	0.2	0.5	1

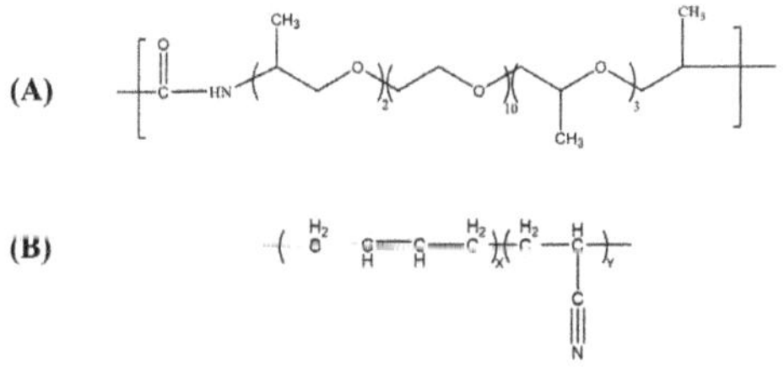

Figure 1. The structural formulas of (**A**) PAE of (**B**) NBR.

The volume of the internal mixer was 300 mL. We first controlled the rotor speed of the mixer to be 60 rpm and the temperature to be 200 °C. NBR, PAE, and antioxidant were added to the mixer for 2 min to prepare the NBR/PAE masterbatch. Then, NBR/PAE masterbatch, N330, SWCNT, zinc oxide, and stearic acid were mixed in the internal mixer for 5 min at 80 °C. Finally, the roll temperature of the two-roll mill was controlled to be 30 °C. The NBR/PAE blend was firstly mixed by wrapping the rolls for 2 min, and then sulfur, accelerator DM, and TMTD were added to the two-roll mill and mixed for 8 min to make sure that the fillers were uniformly dispersed within the blend. NBR/PAE/SWCNT vulcanized rubber was prepared on a flat vulcanizer of 10 MPa at 150 °C. The preparation progress of the NBR/PAE/SWCNT composites is shown in Figure 2.

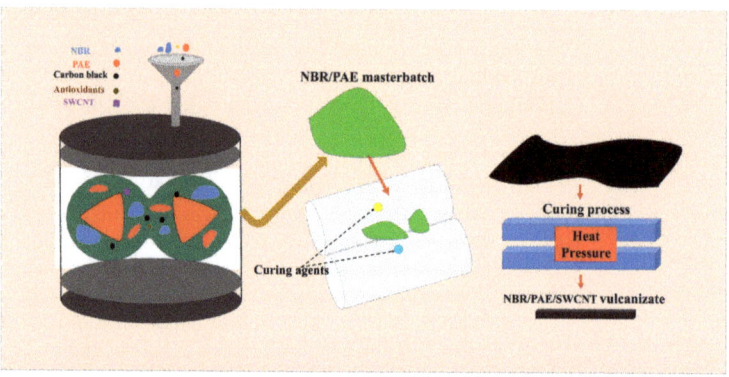

Figure 2. Schematic of the preparation of NBR/PAE/SWCNT composites.

2.2. Experiments

Curing Characteristics. The curing characteristics, such as optimum curing time (t_{90}), scorch time (t_{s2}), minimum torque (M_L), and maximum torque (M_H), were all monitored using the Moving Die Rheometer 2000 (MDR 2000) (Akron, AL, USA) at 150 °C, according to ASTM D2084-95 [35].

Rheological behavior test. The rheological properties of the uncured and cured NBR/PAE/SWCNT composites were investigated using a Rubber Processing Analyzer (RPA 2000, Alpha Technologies, Akron, AL, USA), according to ASTM D6601 [36]. The strain sweep test was performed at 60 °C with 1 Hz frequency in the range of 0.1 to 100% strain. Scanning electron microscopy (SEM) (JSM-6700F, Japan Electronics Corp. Tokyo, Japan). To understand the morphology of the NBR/PAE/SWCNT composites, cured sheets underwent cryogenic fracture. The SEM images were recorded at an accelerating voltage of 8.0 kV.

Tensile test. The tensile testing of the NBR/PAE/SWCNT composites was conducted using a universal testing machine (Zwick/Roell Z005, Esslingen, Germany), according to ASTM D412 [37]. The samples used in the testing were dumbbell-shaped specimens. The gage length of the samples was 30 mm, and the loading speed was 500 mm/min.

Swelling index. The swelling properties of the NBR/PAE/SWCNT composites with different SWCNT contents were assessed by submerging the dry composites in cyclohexanone at room temperature and in IRM 903 and biodiesel at 115 and 135 °C for a set time.

The swelling index of all kinds of samples was calculated according to Equation (1):

$$Swelling\ index = \frac{M_f - M_i}{M_i} \times 100 \qquad (1)$$

where M_i and M_f are the masses of samples before and after immersion in cyclohexanone, IRM 903, and biodiesel, respectively.

Aging process. The thermal aging process was performed at 115 and 135 °C for up to 5 days, according to ASTM D573-04 [38] The specimen was hung directly in the oven (in air) or soaked in biodiesel or IRM 903, then removed when it was finished aging and wiped for testing.

3. Results and Discussion

3.1. Curing Characteristics of the NBR/PAE/SWCNT Composites

Figure 3A shows the vulcanization curves of the NBR/PAE/SWCNT composites with different SWCNT contents. Furthermore, Table 2 summarizes the vulcanization characteristics. It is obvious that after the vulcanization induction period, the torque of all of the samples increased gradually with time. This was due to the network formed by interactions between the rubber or between the rubber and filler. However, at the later stages of vulcanization, the torque decreased slightly. This was likely caused by the thermal cleavage reactions of the crosslinking bonds as well as the chain segments. The addition of SWCNTs caused little change in the NBR/PAE/SWCNT composite vulcanization rates. The maximum (M_H) and minimum torque (M_L) of the NBR/PAE/SWCNT composites increased with the increase in SWCNT content. In general, the crosslinking density of the vulcanized rubber was generally proportional to M_H-M_L. Figure 3B shows that the M_H-M_L values of the NBR/PAE/SWCNT composites increased with higher SWCNT content. This may have been due to the high specific area of SWCNTs resulting in strong interactions with the rubber, leading to a crosslinked network in the NBR/PAE/SWCNT composites.

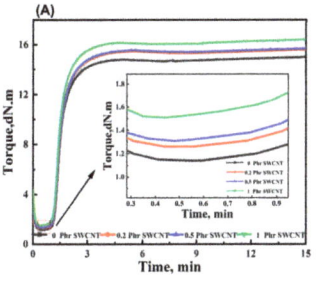
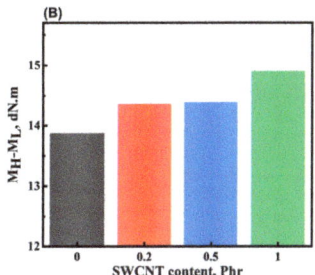

Figure 3. (A) Effect of SWCNT content on curing characteristics of NBR/PAE/SWCNT composites. (B) The M_H-M_L values of NBR/PAE/SWCNT composites.

Table 2. Curing properties of various NBR/PAE/SWCNT composites.

SWCNT Content (phr)	t_{10} (min)	t_{90} (min)	M_L (dNm)	M_H (dNm)
0	1.21	2.51	1.14	15.01
0.2	1.2	2.57	1.26	15.61
0.5	1.21	2.56	1.31	15.70
1	1.19	2.50	1.51	16.41

Note: phr = parts per hundred of rubber.

3.2. RPA 2000 Analysis of the NBR/PAE/SWCNT Composites

The filler network had a significant impact on the properties of the composites. Figure 4A,B shows the storage modulus and loss factor of the uncured NBR/PAE/SWCNT composites versus strain. In summary, the difference between the initial storage modulus (G_0) and the stabilized storage modulus (G') indicated the size of the filler network. As observed in Figure 4A, the filler network of the NBR/PAE/SWCNT composites was gradually enhanced with increased SWCNT content. This could be attributed to the interfacial interactions between SWCNTs and NBR/PAE molecular chains, forming more restricted rubber molecular chains. Moreover, the storage modulus of the composite gradually decreased with strain increase, which was due to destruction of the filler network. Figure 4B shows that the loss factor decreased with increasing SWCNT content at low strains, while the opposite is true at high strains. This was due to the NBR/PAE/SWCNT composites having a strong filler network structure at minor strains. However, at high strains, the filler network was gradually destroyed. The interactions between SWCNTs and NBR/PAE hindered the movement of NBR/PAE molecular chains, which in turn led to an increase in the loss factor.

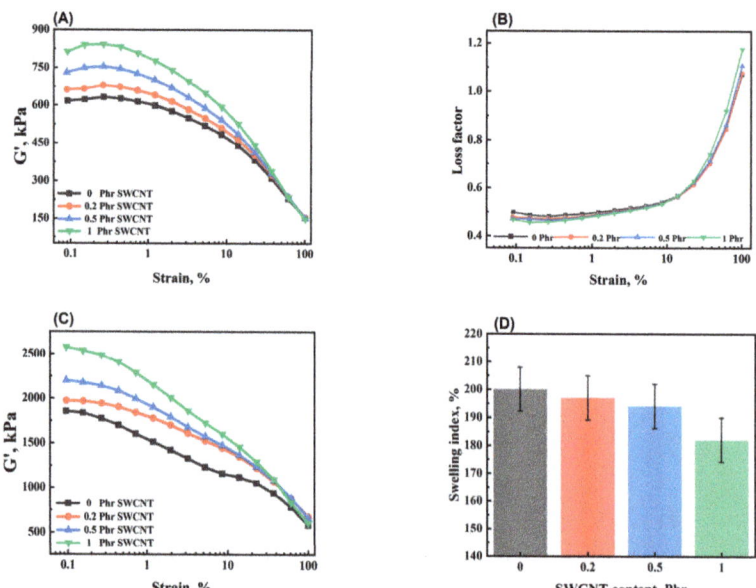

Figure 4. (**A**) Storage modulus (G') and (**B**) loss factor on strain sweep for uncured NBR/PAE/SWCNT composites. (**C**) Storage modulus (G') on strain sweep and (**D**) swelling index for cured composites.

Figure 4C shows the strain scanning curves of vulcanized NBR/PAE/SWCNT composites. The storage modulus gradually decreased with the increase in strain. Moreover, with the increase in SWCNT content, the storage modulus of the NBR/PAE/SWCNT composites increased. This was ascribed to the fact that the interactions and mutual entanglement between SWCNTs and NBR/PAE molecular chains formed physical crosslinking points, increasing the storage modulus. In the next step, the equilibrium swelling method was employed to further investigate the effect of SWCNTs on the NBR/PAE/SWCNT composites. As shown in Figure 4D, the swelling index of the NBR/PAE/SWCNT composites decreased with the increase in SWCNT content. This further proved that SWCNTs could form a crosslinked network with the NBR/PAE molecular chains due to their high aspect ratio and interactions. This limited cyclohexanone entry into the interior of the composites and also correlated well with the results of higher M_H-M_L value, storage modulus, etc.

3.3. Physical Properties and SEM Photographs of the NBR/PAE/SWCNT Composites

The effect of SWCNTs on the mechanical properties of the NBR/PAE/SWCNT composites was studied. Figure 5A shows the stress–strain curves of the NBR/PAE/SWCNT composites with different SWCNT contents. It was found that the 100% constant tensile stress gradually increased with the increase in SWCNT content while the elongation at break progressively decreased. In addition, the hardness of the NBR/PAE/SWCNT composites gradually increased with the increase in SWCNT content, as shown in Figure 5B, which was attributed to the higher aspect ratio and specific surface area of SWCNTs giving a better reinforcing effect to the NBR/PAE/SWCNT composites. This statement was based on Figure 6, which shows SEM images of the NBR/PAE/SWCNT composites. It can be seen that SWCNTs aggregated in the NBR/PAE/SWCNT composites. This may have acted as a stress concentrator, making the composites more susceptible to fracture.

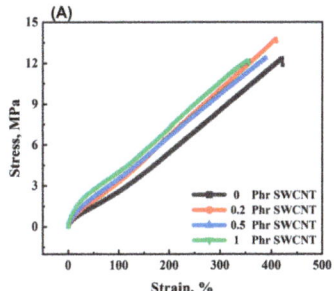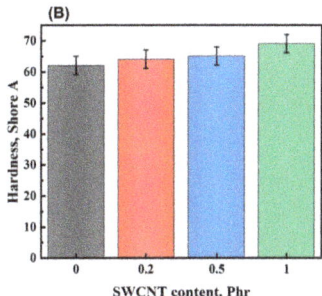

Figure 5. (**A**) Tensile stress–strain curves and (**B**) hardness of NBR/PAE/SWCNT composites.

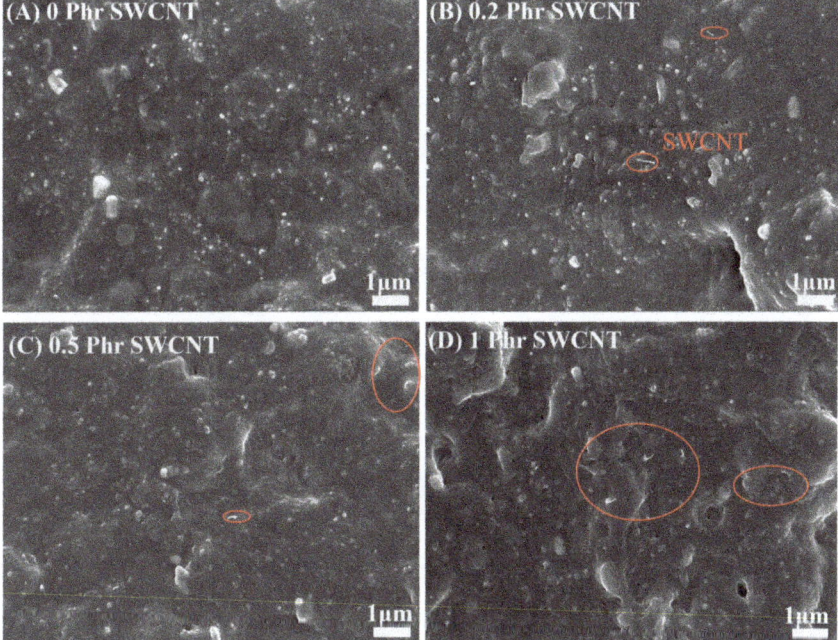

Figure 6. The SEM photographs of (**A**) pure NBR/PAE; (**B**) NBR/PAE-0.2SWCNT; (**C**) NBR/PAE-0.5SWCNT; (**D**) NBR/PAE-1SWCNT.

3.4. The Interactions between SWCNTs and NBR/PAE of the NBR/PAE/SWCNT Composites

The interactions between SWCNTs and NBR/PAE were explored using the Hansen solubility parameter method (HSP) method. The polymer swelling behavior and polymer solvent and polymer–polymer interactions can be predicted using the HSP method [39,40]. Herein, the energy difference (Ra), defined as the spatial distance between polymer and solvent and calculated by Equation (2), was introduced. Three δ^S values refer to the HSPs of the solvent, and three δ^P values refer to the HSPs of the rubber.

$$Ra = \left[4\left(\delta_d^S - \delta_d^P\right)^2 + \left(\delta_p^S - \delta_p^P\right)^2 + \left(\delta_h^S - \delta_h^P\right)^2 \right] \tag{2}$$

The interactions between the rubber and solvent can be expressed in terms of their spatial distance (*Ra*), and *Ra* was calculated to be 3.9 MPa$^{0.5}$ according to Equation (2). In accordance with our previous study, the experimentally-determined radius of the solubility sphere of NBR/PAE was 9.3 MPa$^{0.5}$. The *Ra/Ro* (Figure 7) ratio is called the RED number, which reflects the relative energy difference. RED numbers less than 1.0 indicate high affinity, and progressively higher RED numbers indicate progressively lower affinity [41–43]. The RED between SWCNTs and NBR/PAE was calculated to be 0.42 using Equation (3). This indicated that SWCNTs had better interfacial interactions with NBR/PAE, which further limited the swelling of the NBR/PAE/SWCNT composites. This corresponded to the results of the swelling experiments.

$$RED = \frac{Ra}{Ro} \quad (3)$$

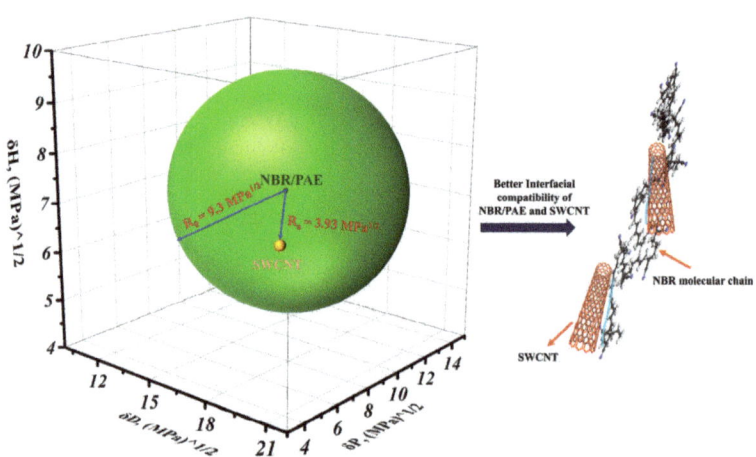

Figure 7. Schematic diagram of the interactions between SWCNTs and NBR/PAE.

Considering the above test results, the reinforcement mechanism of the SWCNT-reinforced NBR/PAE/SWCNT composites is proposed in Figure 8. Because of the higher aspect ratio and specific surface area of SWCNTs, the interactions between SWCNTs and the NBR/PAE matrix resulted in better reinforcement.

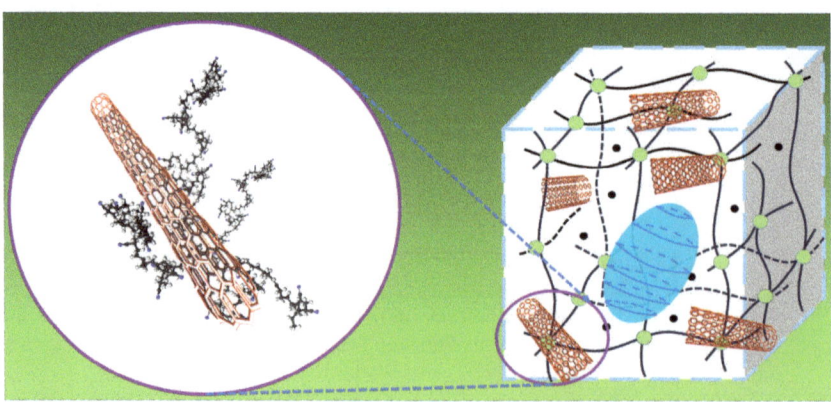

Figure 8. The reinforcement mechanism of SWCNT-reinforced NBR/PAE/SWCNT composites.

3.5. Mechanical Properties before and after Thermo-Oxidative Aging of the NBR/PAE/SWCNT Composites

In previous work, it was found that NBR/PAE blends had better tensile strength retention at different aging temperatures than NBR and PAE, which has high potential for application [30]. In order to further optimize the properties, the effect of different SWCNT contents on the thermal-oxidative aging resistance of the NBR/PAE/SWCNT composites was investigated, as shown in Figure 9. It is clear that as the number of SWCNTs increased, the 100% constant tensile stress increased. As the aging process continued, the elongation at break decreased and 100% constant tensile stress gradually increased. This was attributed to the residual sulfide system not completing the reaction; in the aging process, the molecular chain continued to crosslink, resulting in a decrease in the elongation at break as well as an increase in the 100% constant tensile stress. However, it is interesting to note that the tensile strength of the NBR/PAE/SWCNT composites did not change significantly before and after aging. All of them had excellent tensile strength retention rates. This was attributed to the strong intermolecular interaction between NBR and PAE in the NBR/PAE/SWCNT composites, which was able to resist thermo-oxidative aging. Moreover, SWCNTs could interact with NBR/PAE, which may have led to the high tensile strength retention rates of the NBR/PAE/SWCNT composites under thermo-oxidative aging conditions.

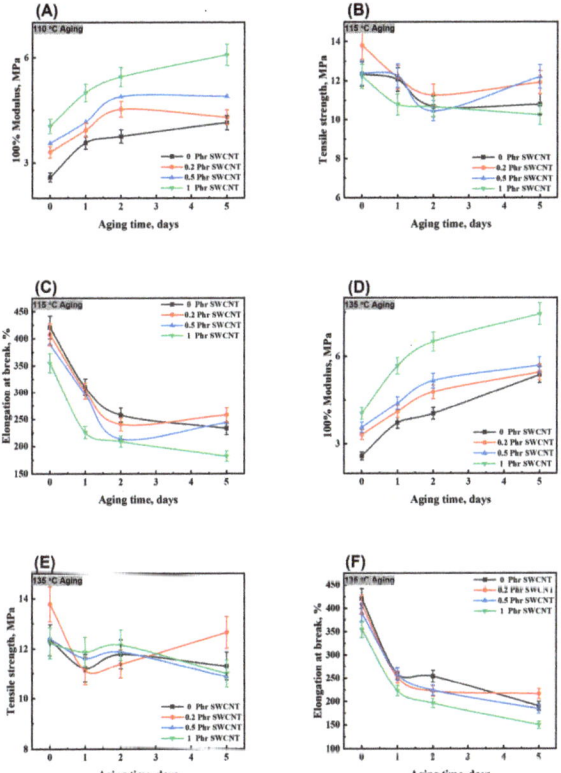

Figure 9. The 100% modulus, tensile strength, and elongation at break of NBR/PAE/SWCNT composites aging at (**A–C**) 115 °C and (**D–F**) 135 °C.

3.6. Properties before and after Oil Aging and the Swelling Behavior of the NBR/PAE/SWCNT Composites

To investigate the changes in NBR/PAE/SWCNT composite properties under oil aging, the NBR/PAE/SWCNT composites were immersed in IRM 903 as well as biodiesel under different aging conditions. After aging in oil for 5 days, the swelling index of the NBR/PAE/SWCNT composites immersed in biodiesel was significantly higher than that in IRM 903, as shown in Figure 10. According to a previous study [44], the HSP method was used to explore the swelling behavior of the NBR/PAE/SWCNT composites in different oils. Equation (4) calculated that the interaction parameters between the NBR/PAE/SWCNT composites and biodiesel were smaller than those with IRM 903 in the aging environment of 115 °C, leading to biodiesel being more likely to enter the composites. As shown in Figures 11 and 12, the tensile strength and elongation at break of the composites after being immersed in the two oils for different aging times were dramatically reduced compared to the unaged samples, and the mechanical properties in biodiesel were more remarkably reduced. This was attributed to the fact that biodiesel entered the rubber more easily and destroyed the molecular chain structure of the rubber, resulting in a greater reduction in tensile strength and elongation at break. Interestingly, the 100% constant tensile stress of the composites immersed in biodiesel gradually decreased with increasing aging time, while the stress gradually increased in IRM 903 oil. After aging in oil at 115 °C, there was a residual sulfide system within the polymer, and the polymer continued to crosslink, increasing the constant elongation stress. Of course, the entry of small molecules into the oil also led to the destruction of the crosslinked polymer network, which resulted in a decrease in the constant tensile stress [30]. When aging in IRM 903 oil, fewer small molecules of oil entered the material compared to that immersed in biodiesel, and the increase in constant stress was due to the post-sulfurization process dominating the process. In contrast, the small molecules of oil had less effect on the disruption of the crosslinked network.

$$\chi_{HSP} = \frac{V_S}{4RT} \times Ra^2 \tag{4}$$

where V_S is the molar volume of the fuel; and R and T are the ideal gas constant and absolute temperature, respectively.

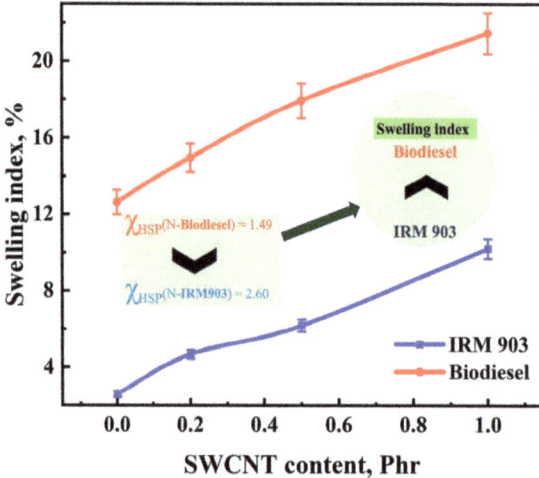

Figure 10. The swelling index of NBR/PAE/SWCNT composites immersed in biodiesel and IRM 903 at 115 °C.

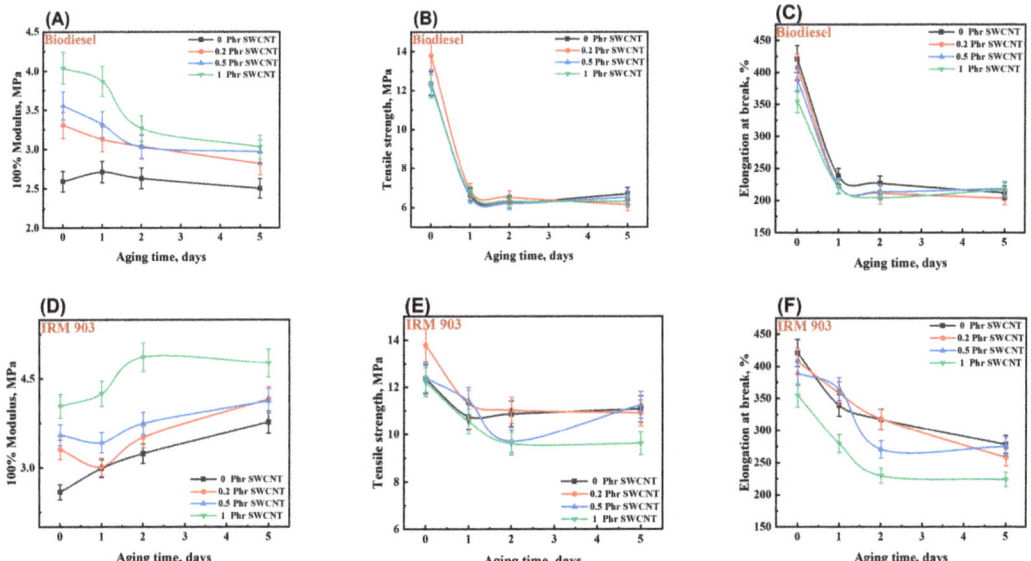

Figure 11. The tensile properties of NBR/PAE/SWCNT composites after immersion in biodiesel (**A–C**) and IRM 903 (**D–F**) for different times at 115 °C.

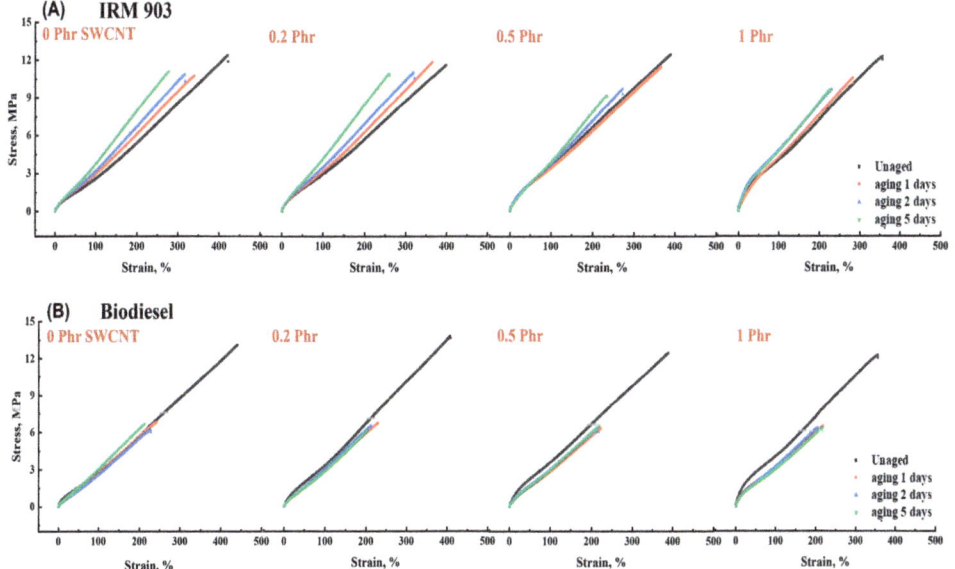

Figure 12. The tensile stress–strain curves of NBR/PAE/SWCNT composites after immersion in IRM 903 (**A**) and biodiesel (**B**) for different times at 115 °C.

Figure 13 shows that the swelling index increased with increasing SWCNT content after aging in two different oils for 5 days. Interestingly, after 5 days of immersion in cyclohexanone, the swelling index decreased with the increase in SWCNT content, showing a different regular variation. Based on this, the relationship between the swelling index and interaction parameters of the NBR/PAE/SWCNT composites with different SWCNT contents within different kinds of oils and solvents was investigated using the HSP method. In some solvents with small interaction parameters, the solvent entered the polymer network more easily, and the crosslinked network swelled to its maximum value. In turn, the interaction forces between SWCNTs and NBR/PAE molecular chains caused the network to retract, decreasing the swelling ratio with increasing SWCNT content. More interestingly, for some oils with large intermolecular interaction parameters, it was difficult for oil to enter the polymer network, and the swelling of the crosslinked network was not significant. Moreover, some weak interfaces between SWCNTs and the rubber were produced, leading to some oil molecules entering more easily. As a result, in some oils with large interaction parameters, the swelling index of the composites increased with increasing SWCNT content. Based on this, the variation in crosslinked polymer networks within liquids with different interaction parameters is plotted in Figure 14, which can better help us to understand the aging process.

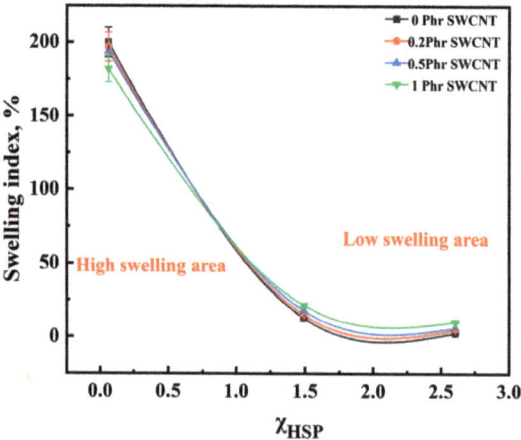

Figure 13. Correlation of the swelling index of NBR/PAE/SWCNT composites with χ_{HSP}.

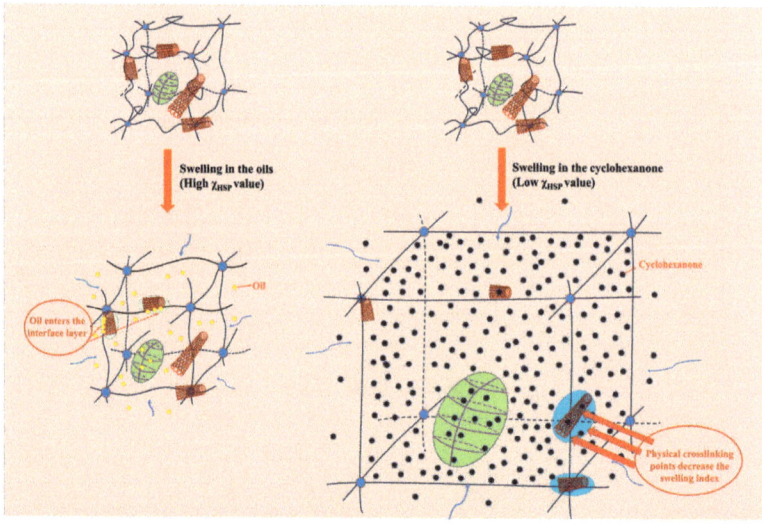

Figure 14. Schematic representation of the changes in the crosslinked network of NBR/PAE/SWCNT immersed in fluids with different interaction parameters.

4. Conclusions

In this work, the interactions between NBR/PAE and SWCNTs were investigated, and the high-temperature aging resistance as well as the oil aging resistance of the composites were studied. The vulcanization curves showed that the $M_H - M_L$ values of the composites gradually increased with the addition of SWCNTs, which was due to their large specific surface area and aspect ratio. The rheological properties of the composites were tested, and the composites storage moduli increased with the SWCNT increase. The 100% constant tensile stress and hardness both increased with the addition of SWCNTs, which was also related to the better reinforcing effect of SWCNTs. The decrease in the elongation at break might have resulted from the aggregation of SWCNTs. Combined with the equilibrium swelling test and analysis by the HSP method, it could be inferred that the interactions between SWCNTs and NBR/PAE formed a better interfacial layer. Thermo-oxidative aging tests at 115 and 135 °C showed that the addition of SWCNTs gave the material higher tensile strength retention and better thermo-oxidative aging properties. Considering that NBR composites are often applied in fields such as the automotive field, contact with oil is unavoidable. Therefore, the composites were considered when immersed in biodiesel at 115 °C and IRM 903 for aging performance testing. The tensile strength of the composites showed different degrees of reduction after aging in the two oils. The 100% constant tensile stress gradually decreased with time in biodiesel but gradually increased in IRM 903.

Interestingly, the swelling index of the rubber material gradually decreased with increasing SWCNT content after immersion in cyclohexanone for 5 days. However, the opposite phenomenon occurred after immersion in biodiesel as well as IRM 903. It could be calculated using the HSP method that the interaction parameters between the composite and cyclohexanone were much smaller than those with biodiesel and IRM 903, which led to a higher swelling index in cyclohexanone. The reason for the different swelling behaviors in different liquids could be deduced from the interaction parameters calculated using the HSP method and combined with the swelling index. The swelling volume of the rubber was regulated by controlling the swelling behavior. This is more favorable for some applications in which volume swelling is required. For example, some rubber sealings need swelling to a certain extent in order to achieve better sealing performance. In summary, this work provides good direction for the development and application of oil- and high-temperature-resistant products. Such products would be suitable for a variety of environments, providing protect against grease, fuel, and chemicals. Next, the abrasion resistance as well as chemical stability of the rubber in liquid environments need to be further explored.

Author Contributions: Conceptualization, G.L. and S.L.; Methodology, H.W.; Software, H.W.; Validation, G.L.; Formal analysis, H.W.; Investigation, T.R.; Resources, S.L.; Data curation, Y.C.; Writing—original draft, H.W.; Writing—review & editing, H.W. and S.L. All authors have read and agreed to the published version of the manuscript.

Funding: This work was supported by the Shandong Provincial Natural Science Foundation (ZR2020QE079).

Institutional Review Board Statement: Not applicable.

Data Availability Statement: The data presented in this study are available on request from the corresponding author.

Conflicts of Interest: The authors declare no conflict of interest.

References

1. Bhaumik, S.; Kumaraswamy, A.; Guruprasad, S.; Bhandari, P. Investigation of friction in rectangular Nitrile-Butadiene Rubber (NBR) hydraulic rod seals for defence applications. *J. Mech. Sci. Technol.* **2015**, *29*, 4793–4799. [CrossRef]
2. Dharmaraj, M.M.; Chakraborty, B.C.; Begum, S. The effect of graphene and nanoclay on properties of nitrile rubber/polyvinyl chloride blend with a potential approach in shock and vibration damping applications. *Iran. Polym. J.* **2022**, *31*, 1129–1145. [CrossRef]

3. Degrange, J.M.; Thomine, M.; Kapsa, P.; Pelletier, J.; Chazeau, L.; Vigier, G.; Dudragne, G.; Guerbé, L. Influence of viscoelasticity on the tribological behaviour of carbon black filled nitrile rubber (NBR) for lip seal application. *Wear* **2005**, *259*, 684–692. [CrossRef]
4. Balasooriya, W.; Schrittesser, B.; Wang, C.; Hausberger, A.; Pinter, G.; Schwarz, T. Tribological Behavior of HNBR in Oil and Gas Field Applications. *Lubricants* **2018**, *6*, 20. [CrossRef]
5. Yun, J.; Zolfaghari, A.; Sane, S. Study of hydrogen sulfide effect on acrylonitrile butadiene rubber/hydrogenated acrylonitrile butadiene rubber for sealing application in oil and gas industry. *J. Appl. Polym. Sci.* **2022**, *139*, e52695. [CrossRef]
6. Hoontrakul, P.; Szamosi, J.; Tobing, S.D. Application of nitrile butadiene rubber for flexible, chemically protective coating. *Polym. Eng. Sci.* **1988**, *28*, 1052–1055. [CrossRef]
7. Pan, C.; Liu, P. Revisiting the surface olefin cross-metathesis of nitrile butadiene rubber on palygorskite nanorods: Product controlling for specific applications. *Appl. Clay Sci.* **2023**, *231*, 106757. [CrossRef]
8. Munusamy, Y.; Kchaou, M. Usage of eggshell as potential bio-filler for arcylonitrile butadiene rubber (NBR) latex film for glove applications. *Ain Shams Eng. J.* **2023**, *14*, 102512. [CrossRef]
9. Porter, C.; Zaman, B.; Pazur, R. A critical examination of the shelf life of nitrile rubber O-Rings used in aerospace sealing applications. *Polym. Degrad. Stab.* **2022**, *206*, 110199. [CrossRef]
10. Budrugeac, P. Thermooxidative degradation of some nitrile-butadiene rubbers. *Polym. Degrad. Stab.* **1992**, *38*, 165–172. [CrossRef]
11. Coronado, M.; Montero, G.; Valdez, B.; Stoytcheva, M.; Eliezer, A.; García, C.; Campbell, H.; Pérez, A. Degradation of nitrile rubber fuel hose by biodiesel use. *Energy* **2014**, *68*, 364–369. [CrossRef]
12. Jiang, B.; Jia, X.; Wang, Z.; Wang, T.; Guo, F.; Wang, Y. Influence of Thermal Aging in Oil on the Friction and Wear Properties of Nitrile Butadiene Rubber. *Tribol. Lett.* **2019**, *67*, 86. [CrossRef]
13. Li, B.; Li, S.-X.; Shen, M.-X.; Xiao, Y.-L.; Zhang, J.; Xiong, G.-Y.; Zhang, Z.-N. Tribological behaviour of acrylonitrile-butadiene rubber under thermal oxidation ageing. *Polym. Test* **2021**, *93*, 106954. [CrossRef]
14. Hota, N.K.; Karna, N.; Dubey, K.A.; Tripathy, D.K.; Sahoo, B.P. Effect of temperature and electron beam irradiation on the dielectric properties and electromagnetic interference shielding effectiveness of ethylene acrylic elastomer/millable polyurethane/SWCNT nanocomposites. *Eur. Polym. J.* **2019**, *112*, 754–765. [CrossRef]
15. Hsiao, F.R.; Wu, I.F.; Liao, Y.C. Porous CNT/rubber composite for resistive pressure sensor. *J. Taiwan. Inst. Chem. Eng.* **2019**, *102*, 387–393. [CrossRef]
16. Yoon, J.; Lee, J.; Hur, J. Stretchable Supercapacitors Based on Carbon Nanotubes-Deposited Rubber Polymer Nanofibers Electrodes with High Tolerance against Strain. *Nanomaterials* **2018**, *8*, 541. [CrossRef] [PubMed]
17. Bakošová, D.; Bakošová, A. Testing of Rubber Composites Reinforced with Carbon Nanotubes. *Polymers* **2022**, *14*, 3039. [CrossRef] [PubMed]
18. Yang, Z.; McElrath, K.; Bahr, J.; D'Souza, N.A. Effect of matrix glass transition on reinforcement efficiency of epoxy-matrix composites with single walled carbon nanotubes, multi-walled carbon nanotubes, carbon nanofibers and graphite. *Compos. Part B Eng.* **2012**, *43*, 2079–2086. [CrossRef]
19. Shafik, E.S.; Tharwat, C.; Abd-El-Messieh, S.L. Utilization study on red brick waste as novel reinforcing and economical filler for acrylonitrile butadiene rubber composite. *Clean. Technol. Environ. Policy.* **2023**, *25*, 1605–1615. [CrossRef]
20. Tang, D.; Zhang, X.; Hu, S.; Liu, X.; Ren, X.; Hu, J.; Feng, Y. The reuse of red brick powder as a filler in styrene-butadiene rubber. *J. Clean. Prod.* **2020**, *261*, 120966. [CrossRef]
21. Zhang, X.; Xue, X.; Yin, Q.; Jia, H.; Wang, J.; Ji, Q.; Xu, Z. Enhanced compatibility and mechanical properties of carboxylated acrylonitrile butadiene rubber/styrene butadiene rubber by using graphene oxide as reinforcing filler. *Compos. Part B Eng.* **2017**, *111*, 243–250. [CrossRef]
22. Ambilkar, S.C.; Das, C. Surface modification of zirconia by various modifiers to investigate its reinforcing efficiency toward nitrile rubber. *Polym. Compos.* **2023**, *44*, 1512–1521. [CrossRef]
23. Utrera-Barrios, S.; Martínez, M.F.; Mas-Giner, I.; Verdejo, R.; López-Manchado, M.A.; Hernández Santana, M. New recyclable and self-healing elastomer composites using waste from toner cartridges. *Compos. Sci. Technol.* **2023**, *244*, 110292. [CrossRef]
24. Wang, F.; Dong, S.; Wang, Z.; He, H.; Huang, X.; Liu, D.; Zhu, H. Self-assembled carbon nanofibers–silica nanocomposites for hydrogenated nitrile butadiene rubber reinforcement. *Polym. Compos.* **2021**, *42*, 5830–5838. [CrossRef]
25. Okuyama, R.; Izumida, W.; Eto, M. Topological classification of the single-wall carbon nanotube. *Phys. Rev. B* **2019**, *99*, 115409. [CrossRef]
26. Ko, J.; Joo, Y. Review of Sorted Metallic Single-Walled Carbon Nanotubes. *Adv. Mater. Interfaces* **2021**, *8*, 2002106. [CrossRef]
27. Cui, J.; Yang, D.; Zeng, X.; Zhou, N.; Liu, H. Recent progress on the structure separation of single-wall carbon nanotubes. *Nanotechnology* **2017**, *28*, 452001. [CrossRef]
28. Kobashi, K.; Ata, S.; Yamada, T.; Futaba, D.N.; Hata, K. Controlling the structure of arborescent carbon nanotube networks for advanced rubber composites. *Compos. Sci. Technol.* **2018**, *163*, 10–17. [CrossRef]
29. Gumede, J.I.; Carson, J.; Hlangothi, S.P.; Bolo, L.L. Effect of single-walled carbon nanotubes on the cure and mechanical properties of reclaimed rubber/natural rubber blends. *Mater. Today Commun.* **2020**, *23*, 100852. [CrossRef]
30. Wang, H.; Liu, S.; Liu, G. Investigation on the thermo-oxidative aging resistance of nitrile-butadiene rubber/polyamide elastomer blend and the swelling behaviors in fuels predicted via Hansen solubility parameter method. *Polym. Degrad. Stab.* **2023**, *217*, 110512. [CrossRef]

31. Mei, S.; Wang, J.; Wan, J.; Wu, X. Preparation Methods and Properties of CNT/CF/G Carbon-Based Nano-Conductive Silicone Rubber. *Appl. Sci.* **2023**, *13*, 6726. [CrossRef]
32. Shahamatifard, F.; Rodrigue, D.; Mighri, F. Thermal and mechanical properties of carbon-based rubber nanocomposites: A review. *Plast. Rubber Compos.* **2023**, *52*, 483–505. [CrossRef]
33. Kitisavetjit, W.; Nakaramontri, Y.; Pichaiyut, S.; Wisunthorn, S.; Nakason, C.; Kiatkamjornwong, S. Influences of carbon nanotubes and graphite hybrid filler on properties of natural rubber nanocomposites. *Polym. Test.* **2021**, *93*, 106981. [CrossRef]
34. Yan, G.; Han, D.; Li, W.; Qiu, J.; Jiang, C.; Li, L.; Wang, C. Effect of pyrolysis carbon black and carbon nanotubes on properties of natural rubber conductive composites. *J. Appl. Polym. Sci.* **2022**, *139*, 52321. [CrossRef]
35. *ASTM D2084-95*; Standard Test Method for Rubber Property-Vulcanization Using Oscillating Disk Cure Meter. American Society for Testing and Materials: West Conshohocken, PA, USA, 2019.
36. *ASTM D6601*; Standard Test Method for Rubber Properties—Measurement of Cure and After—Cure Dynamic Properties Using a Rotorless Shear Rheometer. American Society for Testing and Materials: West Conshohocken, PA, USA, 2019.
37. *ASTM D412*; Standard Test Methods for Vulcanized Rubber and Thermoplastic Elastomer Tension. American Society for Testing and Materials: West Conshohocken, PA, USA, 2022.
38. *ASTM D573-04*; Standard Test Method for Rubber—Deterioration in an Air Oven. American Society for Testing and Materials: West Conshohocken, PA, USA, 2010.
39. Jiang, X.; Hao, Y.; Wang, H.; Tu, J.; Liu, G. Application of Three-Dimensional Solubility Parameter in Diffusion Behavior of Rubber-Solvent System and Its Predictive Power in Calculating the Key Parameters. *Macromol. Res.* **2022**, *30*, 271–278. [CrossRef]
40. Liu, S.S.; Li, X.P.; Qi, P.J.; Song, Z.J.; Zhang, Z.; Wang, K.; Qiu, G.X.; Liu, G.Y. Determination of three-dimensional solubility parameters of styrene butadiene rubber and the potential application in tire tread formula design. *Polym. Test.* **2020**, *81*, 106170. [CrossRef]
41. Otárola-Sepúlveda, J.; Cea-Klapp, E.; Aravena, P.; Ormazábal-Latorre, S.; Canales, R.I.; Garrido, J.M.; Valerio, O. Assessment of Hansen solubility parameters in deep eutectic solvents for solubility predictions. *J. Mol. Liq.* **2023**, *388*, 122669. [CrossRef]
42. Li, M.; Ren, T.; Sun, Y.; Xiao, S.; Wang, Y.; Lu, M.; Zhang, S.; Du, K. New Parameter Derived from the Hansen Solubility Parameter Used to Evaluate the Solubility of Asphaltene in Solvent. *ACS Omega* **2022**, *7*, 13801–13807. [CrossRef]
43. Larson, B.K.; Hess, J.M.; Williams, J.M., II. Procedure for estimating oil three-dimensional solubility parameters. *Rubber Chem. Technol.* **2017**, *90*, 621–632. [CrossRef]
44. Su, R.; Liu, G.; Sun, H.; Yong, Z. A new method to measure the three-dimensional solubility parameters of acrylate rubber and predict its oil resistance. *Polym. Bull.* **2022**, *79*, 971–984. [CrossRef]

Disclaimer/Publisher's Note: The statements, opinions and data contained in all publications are solely those of the individual author(s) and contributor(s) and not of MDPI and/or the editor(s). MDPI and/or the editor(s) disclaim responsibility for any injury to people or property resulting from any ideas, methods, instructions or products referred to in the content.

Article

Flexible Composite Electrolyte Membranes with Fast Ion Transport Channels for Solid-State Lithium Batteries

Xiaojun Ma [1], Dongxu Mao [1], Wenkai Xin [1], Shangyun Yang [1], Hao Zhang [1], Yanzhu Zhang [1], Xundao Liu [1], Dehua Dong [2], Zhengmao Ye [1,*] and Jiajie Li [1,*]

[1] School of Materials Science and Engineering, University of Jinan, Jinan 250022, China; 15552600076@163.com (X.M.); 17663061976@163.com (D.M.); xinwenkai2021@163.com (W.X.); 17860612220@163.com (S.Y.); 18669576201@163.com (H.Z.); 18231895913@163.com (Y.Z.); mse_liuxundao@ujn.edu.cn (X.L.)

[2] Department of Chemical and Biological Engineering, Monash University, Clayton, VIC 3800, Australia; dehua.dong@monash.edu

* Correspondence: mse_yezm@ujn.edu.cn (Z.Y.); mse_lijj@ujn.edu.cn (J.L.); Tel.: +86-531-89736652 (Z.Y.); Fax: +86-531-89736012 (J.L.)

Abstract: Numerous endeavors have been dedicated to the development of composite polymer electrolyte (CPE) membranes for all-solid-state batteries (SSBs). However, insufficient ionic conductivity and mechanical properties still pose great challenges in practical applications. In this study, a flexible composite electrolyte membrane (FCPE) with fast ion transport channels was prepared using a phase conversion process combined with in situ polymerization. The polyvinylidene fluoride-hexafluoro propylene (PVDF-HFP) polymer matrix incorporated with lithium lanthanum zirconate (LLZTO) formed a 3D net-like structure, and the in situ polymerized polyvinyl ethylene carbonate (PVEC) enhanced the interface connection. This 3D network, with multiple rapid pathways for Li$^+$ that effectively control Li$^+$ flux, led to uniform lithium deposition. Moreover, the symmetrical lithium cells that used FCPE exhibited high stability after 1200 h of cycling at 0.1 mA cm^{-2}. Specifically, all-solid-state lithium batteries coupled with LiFePO$_4$ cathodes can stably cycle for over 100 cycles at room temperature with high Coulombic efficiencies. Furthermore, after 100 cycles, the infrared spectrum shows that the structure of FCPE remains stable. This work demonstrates a novel insight for designing a flexible composite electrolyte for highly safe SSBs.

Keywords: PVDF-HFP/LLZTO; net-like structure; flexible composite electrolyte; PVEC

Citation: Ma, X.; Mao, D.; Xin, W.; Yang, S.; Zhang, H.; Zhang, Y.; Liu, X.; Dong, D.; Ye, Z.; Li, J. Flexible Composite Electrolyte Membranes with Fast Ion Transport Channels for Solid-State Lithium Batteries. *Polymers* **2024**, *16*, 565. https://doi.org/10.3390/polym16050565

Academic Editors: Yuwei Chen and Yumin Xia

Received: 8 January 2024
Revised: 11 February 2024
Accepted: 15 February 2024
Published: 20 February 2024

Copyright: © 2024 by the authors. Licensee MDPI, Basel, Switzerland. This article is an open access article distributed under the terms and conditions of the Creative Commons Attribution (CC BY) license (https:// creativecommons.org/licenses/by/ 4.0/).

1. Introduction

The escalating demand for highly secure energy storage systems in wearable electronics and power batteries has underscored the imperative for advancements in lithium batteries [1–5]. SSBs with higher safety can be a promising candidate for a spectrum of issues associated with traditional liquid electrolyte systems [6–9].

Solid electrolytes are broadly categorized into two primary classes: organic and inorganic [10–12]. Generally, inorganic electrolytes, exemplified by oxides such as Li$_7$La$_3$Zr$_2$O$_{12}$ (LLZO), exhibit notable mechanical strength and commendable thermal stability, while the ionic conductivity exceeds 10^{-3} S cm^{-1} at room temperature [13–15]. Despite these merits, the inherent rigidity of inorganic electrolytes poses challenges at the electrode–electrolyte interface, resulting in elevated interface resistance [16]. Conversely, polymer electrolytes, typified by materials like polyethylene oxide and polyvinylidene fluoride (PVDF), establish favorable interfacial contacts with diverse electrodes [17]. However, their lithium-ion conductivity at room temperature, approximately ~10^{-6} S cm^{-1}, is comparatively low, hindering the advancement of organic polymer electrolytes to a certain extent [18–21].

A novel approach involves incorporating an inorganic electrolyte as an active filler into an organic polymer electrolyte to formulate a polymer composite solid electrolyte [22,23].

This strategy leverages the flexibility inherent in solid electrolytes, resulting in an improvement in ionic conductivity at room temperature [24]. Recent investigations have demonstrated progress in the development of composite polymer electrolytes (CPE), particularly those incorporating LLZO as an active filler [25]. Such formulations exhibit superior mechanical energy characteristics and enhanced ionic conductivity at room temperature compared to other composite polymer electrolytes of similar compositions [26–28].

The method for preparing porous electrolytes typically involves the use of extrusion and casting, resulting in porous electrolyte films through solvent evaporation [29]. Y.L et al. [30] prepared a composite solid electrolyte using a solution casting method. However, the pores obtained using this method are often not sufficiently interconnected and have relatively small diameters, making it difficult to store electrolyte materials within these pores. As a result, the electrolyte is prone to consumption, leading to insufficient stability in cycling performance. In recent years, phase transition methods have gained research significance as a pore-forming technique [31]. This method involves a polymer solution system where the solvent is the continuous phase, undergoing a process that transforms into a swollen solid state. In simple terms, the polymer dissolves in a solvent, solvent A, to form a homogeneous solution. When this solution is mixed with another solvent, solvent B, which is miscible with A but does not dissolve the polymer, phase separation and solidification occur within the homogeneous solution. If the phase separation occurs rapidly, solvent A within the homogeneous solution is quickly replaced by solvent B, resulting in the formation of pores within the solidified polymer [31–33].

J.Z et al. [34] utilized the phase transition method to prepare a composite solid electrolyte with vertical microchannels. This electrolyte had a significant thickness, leading to longer lithium ion transport distances and increased impedance. By reducing the thickness of the electrolyte film, irregular three-dimensional interconnected large pores were obtained using the phase transition method, creating abundant polymer-active filler interfaces. Additionally, these interconnected large pores provided storage for gel electrolytes, further accelerating lithium ion conduction in a three-dimensional space [35].

In these CPEs, lithium-ion (Li$^+$) transport primarily occurs through a unique pathway within the amorphous region of polymer, encompassing the polymer–filler interface and the active filler [36]. It is always lead to an insignificant increase in ionic conductivity by simply mixing the polymer matrix with ceramic particles due to particle agglomeration and lack of a well-defined ceramic–polymer interface [35,37–39]. The isolated ceramic particles within polymer fail to create a lithium-ion conductive network, resulting in low conductivity. Recent studies show that constructing a three-dimensional (3D) net-like structure with LLZO filler can reduce particle agglomeration [40]. However, a substantial improvement in ionic conductivity close to $\sim 10^{-4}$ S cm^{-2} at room temperature still remains challenging [41]. To address these issues, establishing a 3D packing network through in situ polymerization within the polymer electrolyte is crucial, which ensures continuous lithium-ion conductivity throughout the structure [42,43]. Thus, precisely optimizing the morphology and content of active ceramic fillers plays a crucial role in enhancing the electrochemical and mechanical properties of CPE [44].

Herein, a flexible composite electrolyte composed of pores 3D net-like structure and fast ion channels was developed. The polyvinylidene fluoride-hexafluoro propylene (PVDF-HFP) and lithium lanthanum zirconate (LLZTO, Li$_{6.4}$Al$_{0.2}$La$_3$Zr$_{1.4}$Ta$_{0.6}$O$_{12}$) based pores 3D net-like structure created a well-defined ceramic–polymer interface which give the electrolyte high mechanical strength and improved ion conductivity. The filled PVEC-based electrolyte in the pores can collect ions on the porous walls, further providing a fast ion migration channel which further enhances the ionic conductivity at room temperature. This approach, permeating ionic conductive species within a three-dimensional network, demonstrates superior electrode–electrolyte interface contact compared to conventional composite solid-state electrolytes. The as prepared FCPE exhibits a high ionic conductivity of about 1.21×10^{-4} S cm^{-1} and cycle stability for 1200 h for a lithium-symmetrical battery at a current density of 0.1 mA cm^{-2}. Importantly, when applied in an all-solid-state

Li | FCPE | LiFePO$_4$ coin cell, it maintains a high specific capacity of 148.5 mA h g^{-1}, even after 100 cycles, demonstrating robust cyclic stability. Furthermore, characterization of the solid electrolyte membrane after 100 cycles and 1200 h using FT-IR spectroscopy revealed that the structure of membrane remains stable. This underscores significant prospects for practical applications in the field of solid-state electrolytes.

2. Materials and Methods

2.1. Materials

Dimethyl sulfoxide (DMSO, >99.9%), Lithium bis(trifluoromethanesulphonyl)imide (LiTFSI, 99.0%), 4-Vinyl-1,3-dioxolan-2-one (VEC, 99.0%), and 2,2′-Azobis(2-methylpropionitrile) (AIBN, 99.0%) were all purchased from Macklin, Shanghai, China. Li$_{6.4}$Al$_{0.2}$La$_3$Zr$_{1.4}$Ta$_{0.6}$O$_{12}$ solid electrolyte powder (LLZTO), Lithium iron phosphate (LFP, P198-S20), and Carbon nanotube dispersants (CNTs) were purchased from Shenzhen Kejing, Shenzhen, China. Poly (vinylidene fluoride-co-hexafluoropropylene) (PVDF-HFP, KynarFlex2801, Mw = 900,000)

2.2. Materials Synthesis

The phase conversion method was used to the preparation of an FCPE with a porous structure. An appropriate quantity of PVDF-HFP was dissolved in dimethyl sulfoxide (DMSO) and magnetically stirred for 60 min at 60 °C to yield a 10 wt% polymer clear solution. Subsequently, LLZTO powder was added to the solution in varying ratios (x LLZTO/y PVDF-HFP = 0%, 5%, 10%, 15%, denoted as PHxL where x = 0, 5, 10, 15). The resulting mixture was further stirred for 20 min under vacuum conditions to achieve a homogeneously dispersed slurry. The slurry was then cast onto a smooth glass plate at 40 °C, and the resulting membrane was transferred to a vessel filled with ultrapure water. After 5 min, it was removed and dried in a forced-air oven at 60 °C for 12 h, yielding the PVDF-HFP-LLZTO membrane. All membranes were subsequently transferred to a glovebox. The membranes exhibit a thickness of PH10L approximately 19 μm and a diameter of 19 mm.

For the PVEC-based electrolyte precursor, 2.5 M lithium hexafluorophosphate (LiPF$_6$) was added to vinylene carbonate (VEC), along with 0.02 wt% azobisisobutyronitrile (AIBN) as an initiator for polymerization. The solid electrolyte, filled with the PVEC-based electrolyte precursor slurry during the battery loading process into the PHxL is denoted as PHxLE.

In the preparation of the cathode electrode, lithium iron phosphate (LFP), PVDF-HFP and Carbon nanotube dispersants (CNTs) (the mass ratio of the three is 8:1:1) were mixed and stirred in DMSO for 6 h. The final slurry is poured onto an aluminum foil current collector and dried at 100 °C for 24 h. The active material loading on the aluminum foil current collector is about 2.0 mg cm^{-2}.

2.3. Characterization

The morphological features of the specimens were assessed through scanning electron microscopy (SEM) utilizing the ZEISS Sigma 300 instrument. Transmission electron microscopy (TEM) and elemental mapping images were acquired using a transmission electron microscope (JEM-2100P, JEOL, Tokyo, Japan). X-ray diffraction (XRD) patterns were captured on the Smart Lab 9KW instrument with Cu Kα radiation. Thermogravimetric analyses (TGA) were executed employing a PerkinElmer STA 6000 analyzer. Fourier-transform infrared (FT-IR) spectroscopy was tested on a Nicolet 6700 spectrometer. Raman experiments were performed at Horiba Labram Evolution and were measured from 1064 nm power excitation.

2.4. Electrochemical Characterization

Two meticulously polished stainless steel discs (SS) served as blocking electrodes to encapsulate the composite solid electrolyte, establishing a blocking-type cell. The ionic conductivity of FCPE at room temperature (25 °C) was quantified utilizing electrochemical impedance spectroscopy (EIS) in the spectral range of 0.01 to 10^5 Hz at room temperature with an alternating amplitude of 10 mV. The ionic conductivity (σ) has calculated using the equation $\sigma = L/(R_b \cdot S)$.

The electrochemical window of FCPE was determined by employing the linear sweep voltammetry (LSV) technique, which is obtained using a Li/FCPE/SS cell configuration with a scan rate of 10 mV s^{-1} and a scan range of 0 to 4.8 V. Li/FCPE/LiFePO$_4$ coin cells were assembled into all-solid-state batteries, and the LAND CT2001A meter was used to perform a charge–discharge test within the voltage range of 2.5 to 3.65 V at 25 °C.

3. Results and Discussion

As depicted in Figure 1, after the phase conversion process, the solid electrolyte exhibits a porous structure with PVDF-HFP encapsulating ceramic powder particles, forming a rich PVDF-HFP-LLZTO interface, constituting a three-dimensional interconnected channel. This porous structure is impregnated with a PVEC-based electrolyte single-ion conductor through in situ polymerization. At the same time, the VEC has not undergone complete polymerization, and free monomers still exist around the polymer chains. The presence of the VEC monomers leads to a reduction in the crystallinity of the polymer, resulting in a shorter polymer chain length, which further promotes the conduction of lithium ions. Figure 2a shows the as-prepared solid electrolyte has a white appearance and flexible characteristics. SEM images revealed that the porous flexible composite electrolyte possesses an interconnected network structure characterized by nest-shaped polymer ducts that are 3D interconnected (Figure 2b–d). The enlarged view of the red rectangular region selected in Figure 2b is shown in Figure 2c, and the enlarged view of Figure 2c is shown in Figure 2d. This distinctive architecture is attributed to the exchange of solvents and non-solvents during phase conversion, driven by a reduction in the Gibbs free energy of the system through the generation of a new phase (β phase) per unit volume from the parent phase (α phase) [45,46]. Among them, white spots on the surface of the film can be further observed in the red dotted circle in Figure 2d. It is believed that the white spots are ceramic particles wrapped by organic matter. The microstructure of the PHxL 3D skeleton was observed by TEM. As shown in Figure 2e, the skeleton is a porous structure with a pore diameter of approximately 100 nm. It should be noted that nanoparticles with a diameter of about 200 nanometers were embedded within the skeleton which is consistent with the SEM result in Figure 2e. The interface between the LLZTO phase and the elastomer matrix exhibited a smooth connection, elucidating the achieved structural integrity in the composite electrolyte. Further, high-resolution transmission electron microscopy (HRTEM) images (Figure 2f) showed that the surface of pure LLZTO particles was uniformly coated with amorphous PVDF-HFP. The spacing of the lattice fringes of the particles is 0.53 nm, which is in good agreement with the (211) plane of LLZTO, indicating that the PVDF-HFP coating of LLZTO does not destroy the crystal structure of LLZTO during the phase conversion process [47]. It has been reported that a chemical reaction occurs between PVDF-HFP and LLZTO, ultimately leading to the defluorination of PVDF-HFP [48]. Notably, PVDF-HFP exhibits a propensity for defluorination in alkaline environments, a phenomenon exacerbated by the alkaline nature often exhibited by LLZTO ceramics in various solutions. This study employed a phase transition method. During the transition process, PVDF-HFP is insoluble in water and undergoes rapid solvent exchange upon contact. Uniformly dispersed LLZTO particles are encapsulated within PVDF-HFP, forming numerous large interconnected three-dimensional pores.

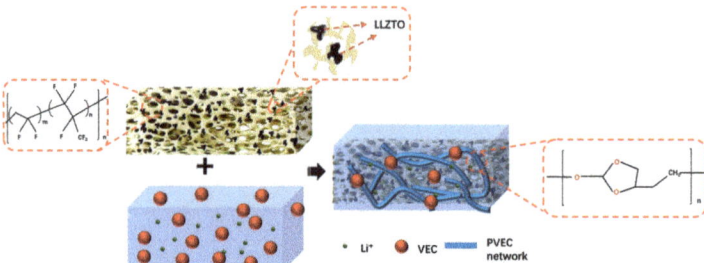

Figure 1. Schematic diagram of the FCPE structure.

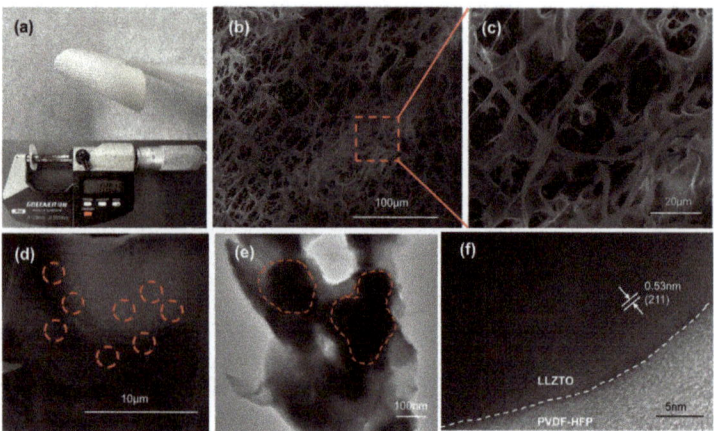

Figure 2. (**a**) Digital image of PH10L; (**b**–**d**) the surface morphology SEM images of PH10L; (**e**) STEM image; and (**f**) high-resolution TEM (HRTEM) image of PH10L.

The XRD pattern of the composite electrolyte is depicted in Figure 3a. XRD peaks of pure PVDF-HFP film at 20° and 40° reveal the presence of polar β-phase and γ-phase [49]. Upon the incorporation of ceramic fillers, the XRD peaks associated with PVDF-HFP weaken and broaden, indicating a reduction in the crystallinity of the PVDF-HFP component in the electrolyte. Additionally, the introduction of LLZTO corresponds to a cubic garnet crystal structure with the Ia3′d space group, matching the standard peaks of $Li_7La_3Nb_2O_{12}$ (PDF#: 40-0894) [50] well. (The PDF profile of the $Li_7La_3Nb_2O_{12}$ which has a similar cubic phase to LLZTO) [35]. The cubic phase of LLZTO is more favorable for lithium-ion conduction compared to the amorphous phase of PVDF-HFP, contributing to the stability of the composite material. Furthermore, with an increase in the LLZTO content, the characteristic peaks associated with LLZTO gradually intensify. The FTIR spectrum of the electrolyte membrane is depicted in Figure 3b. Vibrational peaks at 840, 1234, 1275, and 1423 cm^{-1} corroborate the XRD results by corroborating the β- and γ-phases of PVDF-HFP [51]. In the Raman spectrum (Figure 3c), as the LLZTO content increases, the peaks at 1120 and 1510 cm^{-1} gradually intensify, likely attributed to C-C stretching vibrations. This indicates that the PVDF-HFP backbone has been modified. This suggests that with the incorporation of LLZTO, the main chain structure of PVDF-HFP undergoes dehydrofluorination reactions [52]. Notably, the FT-IR spectrum of PVEC exhibited prominent peaks corresponding to the C=O bond at 1795 cm^{-1} and the C-O bond at 1063 cm^{-1}, indicative of the stability of carbonate units during the thermal polymerization process (Figure S1) [49]. Furthermore, the incorporation of VEC into the framework of PH10L for polymerization results in PH10LE, as shown in Figure 3d. The infrared spectrum of PH10LE exhibits almost identical characteristic peaks to those of PH10L and PVEC, indicating that its structure remains unchanged after the formation of the composite.

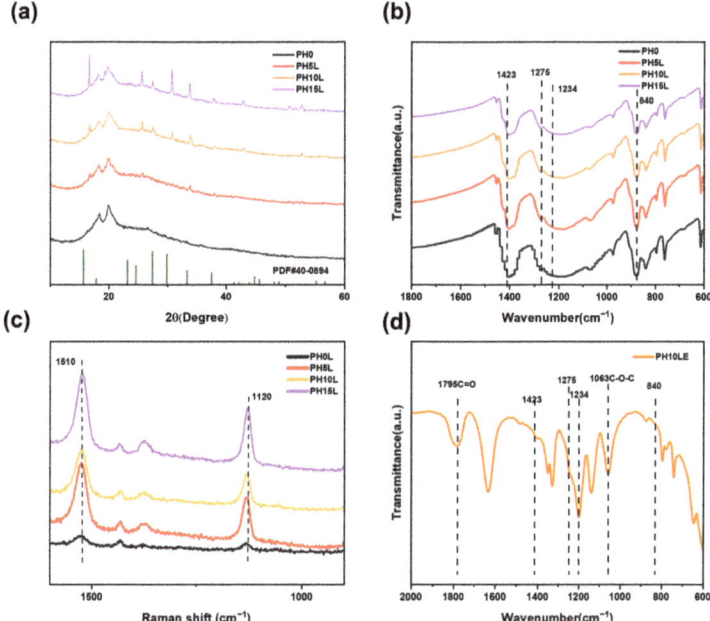

Figure 3. (a) XRD curves, (b) FT-IR spectrum, and (c) Raman spectra of PH0LE to PH15LE at room temperature. (d) FT-IR spectrum of PH10LE at room temperature.

The thermal characteristics of the PH0L and PH10L films were elucidated through TGA. As shown in Figure S2, below 400 °C, both pure PVDF-HFP membrane and PH10L remain stable without decomposition. Both experience gradual weight loss and abrupt decomposition around 450 °C, indicating a decrease in their thermal stability. However, the network-like electrolyte film with LLZTO fillers exhibits a 20 °C higher temperature for rapid thermal decomposition compared to pure PVDF-HFP after the addition of LLZTO fillers, suggesting an improvement in thermal stability. Moreover, at temperatures as high as 550 °C, pure PVDF-HFP completely decomposes, while PH10L still retains undecomposed ceramic particles, demonstrating the beneficial effect of incorporating ceramic powders. TGA curves illustrated the initial thermal decomposition of PH0L at 428 °C, and both films exhibited commendable thermal stability, with no decomposition observed up to 400 °C (Figure S2) [27,35]. Simultaneously, the issue of thermal runaway in lithium batteries is a critically important safety concern. In this study, a combustion test was employed to investigate the flame retardancy of PH10L and a commercially available PE separator. As depicted in Figure 4a, the commercial PE separator was easily ignited, and by the third second, it was completely consumed in flames, with observable dripping of combustion by-products. In contrast, when a flame was introduced near PH10L, the shrinkage rate significantly decreased. If the flame source was removed, PH10L ceased combustion, as illustrated in Figure 4b. In addition, we have also provided video material of the combustion experiments. This suggests that the introduction of LLZTO resulted in a postponement of the separator's combustion. Furthermore, the outstanding thermal stability of PH10L was substantiated by storing the separators at different ambient temperatures. As clearly observed in Figure 4c, a distinct trend emerged. With increasing temperature, the commercially available PE separator exhibited pronounced curling, whereas at 120 °C,

PH10L maintained a high level of flatness. This visually demonstrates the thermal stability of PH10L, consistent with the earlier-discussed TG results and combustion tests.

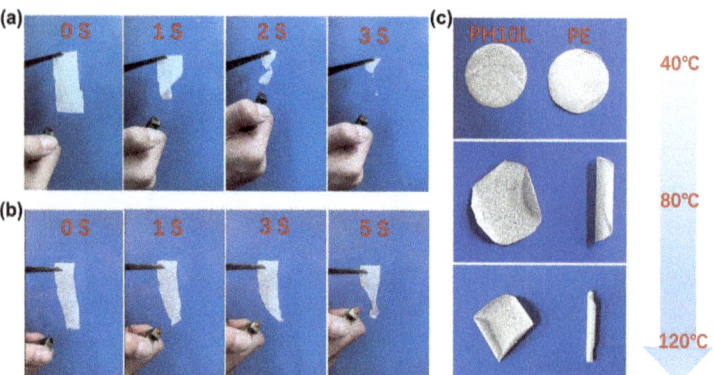

Figure 4. (**a**) Combustion test of commercial PE separator and (**b**) PH10L. (**c**) The digital photos of PH10L and PE separator heated at different temperatures.

Electrochemical Performance

Systematic studies were conducted on the ionic conductivity of solid electrolytes using SS blocking electrodes. Figure 5a displays the EIS impedance spectra of PHxL CPE at room temperature. All curve shapes are consistent with a high-frequency region showing a downward-bent semicircular arc and a low-frequency region corresponding to a sloping straight line. The AC impedance spectra were fitted using an equivalent circuit through least squares fitting, yielding the bulk resistivity of the electrolyte. With the continuous increase in LLZTO content, the lithium-ion conductivity of the flexible electrolyte also increases. The ion conductivity of PH0LE is only 4.0×10^{-5} S cm^{-1}. However, when the LLZTO content reaches 10%, the ion conductivity of PH10LE is maximized, reaching up to 1.21×10^{-4} S cm^{-1}, which is much higher than that of PH0LE. This may be attributed to the addition of ceramic powders affecting the crystallinity of the polymer, resulting in an overall enhancement of the ionic conductivity [50]. Furthermore, the ionic conductivity of PH10LE and PH0LE was further tested at different temperatures, as depicted in the Arrhenius plot in Figure 5b. The temperature range for the FCPE electrolyte was set between 15 °C and 60 °C. From this, it can be observed that with an increase in temperature, the ionic conductivity of FCPE gradually increases. The ion conductivity of composite solid-state electrolytes in all-solid-state lithium-ion batteries without the addition of liquid electrolyte, as presented in Table S1 from recent reports, clearly indicates that the final results obtained in this study are outstanding [45,50]. Combining the analysis with Figure 5c, the increase in the amorphous region is also beneficial for the rapid migration of lithium ions. It elucidates the temporal evolution of current throughout the polarization process. The accompanying diagrams delineate the simulated equivalent circuit and impedance spectra, wherein R$_s$, R$_{ct}$, and R$_2$ signify the resistances arising from the electrolyte bulk, grain boundaries, and both electrode–electrolyte interfaces, respectively. CPE$_1$ and CPE$_2$ are associated with constant phase elements at grain boundaries, LLZTO, and polymer interfaces. W$_s$ characterizes the impedance encountered by reactants diffusing from the electrolyte bulk to the electrode reaction interface. The impedance response was meticulously computed using ZView software [37]. The 10 wt% LLZTO ensures that the amorphous region of the organic material is maximized without particle agglomeration, and the network structure itself is interconnected. At this concentration, 10 wt% LLZTO is sufficient to form a three-dimensional interconnected Li$^+$ channel. Excessive LLZTO, on the other hand, can hinder ion transport, thus reducing the conductivity of Li$^+$ [48]. Within this porous structure saturated with VEC precursors, the polymerized PVEC-based electrolyte serves as an ion conductor, intricately interacting with the 3D porous skeleton. This interaction results in the creation of a unique channel, facilitating the rapid transport of lithium ions. Lithium ions can not only be transported within the 3D net-like structure but also within the PVEC-based electrolyte, further enhancing lithium ion conductivity [51]. As shown in Figure 5d, the LSV measurements were measured with a voltage range of 0 to 4.8 V at room temperature. Compared to the PH0LE, the potential of PH10LE begins to decompose because of oxidation at as high as 4.7 V relative to Li$^+$/Li which indicates that the PH10LE is very stable at high voltage.

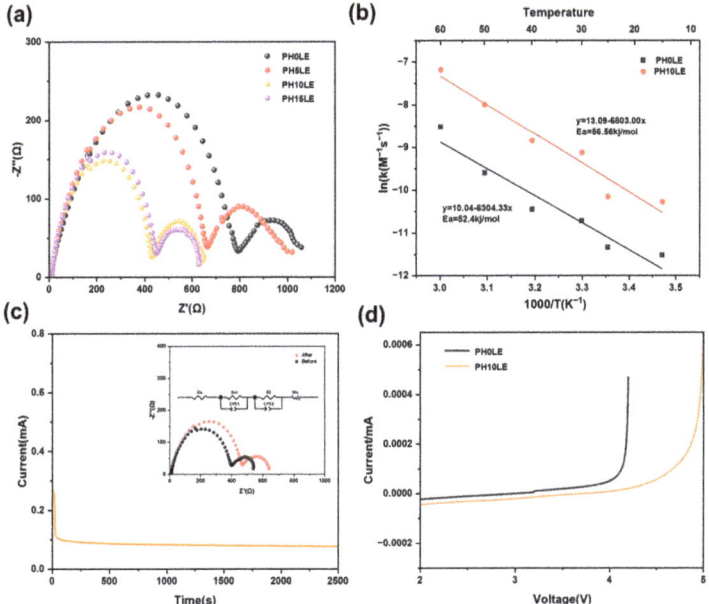

Figure 5. (**a**) EIS profiles of PH0LE to PH15LE, (**b**) The ionic conductivity of the PH10LE and PH0LE separator at different temperatures. (**c**) i-t curves of PH0LE and PH10LE at room temperature. (**d**) LSV of PH0LE and PH10LE at room temperature.

Lithium plating/stripping experiments were conducted in a symmetrical cell to demonstrate the interface stability of FCPE on the lithium metal. As shown in Figure 6a, the voltage-time curve indicates that the battery with PH10LE CPE can maintain stability at a current density of 0.1 mA cm^{-2}, remaining stable even after 1200 h of cycling. In contrast, the battery using PH0LE CPE experiences a short circuit within 600 h (Figure 6a). Furthermore, compared to other lithium batteries, the polarization voltage of the Li|PH10LE|Li lithium battery is only 10 mV. These results suggest that this mesh-like FCPE is more effective in suppressing lithium dendrite growth and relatively regulating lithium deposition.

To further illustrate the advantages of the 3D net-like flexible composite electrolyte, a solid-state lithium metal battery was constructed with LFP as the cathode and its electrochemical performance was tested. In Figure 6b, it depicts the reversible charge–discharge capacities of the lithium metal battery at different rates (0.1 C–0.5 C). The stable capacities of the Li|PH10LE|LFP battery at 0.1 C, 0.2 C, and 0.5 C are 155.8, 146.3, and 136.7 mA h g^{-1}, respectively. Upon switching back to 0.1 C, the capacity can recover to 152.1 mA h g^{-1}, demonstrating outstanding rate capability. Figure 6c compares the long-term cycling performance and corresponding coulombic efficiency of the two batteries at 0.1 C. All-solid-state batteries assembled with PH0LE are shorted directly after the 20th cycle. In contrast, the Li|PH10LE|LFP battery achieves over 100 reversible electrochemical cycles and maintains a high specific capacity of 148.5 mA h g^{-1}. Figure 6d shows the charge–discharge curves of the Li | PH10LE | LFP battery at different cycles at 0.1 C. From the 5th to the 35th cycle, the curves remain smooth, and the capacity stays relatively stable, indicating the absence of side reactions within the battery.

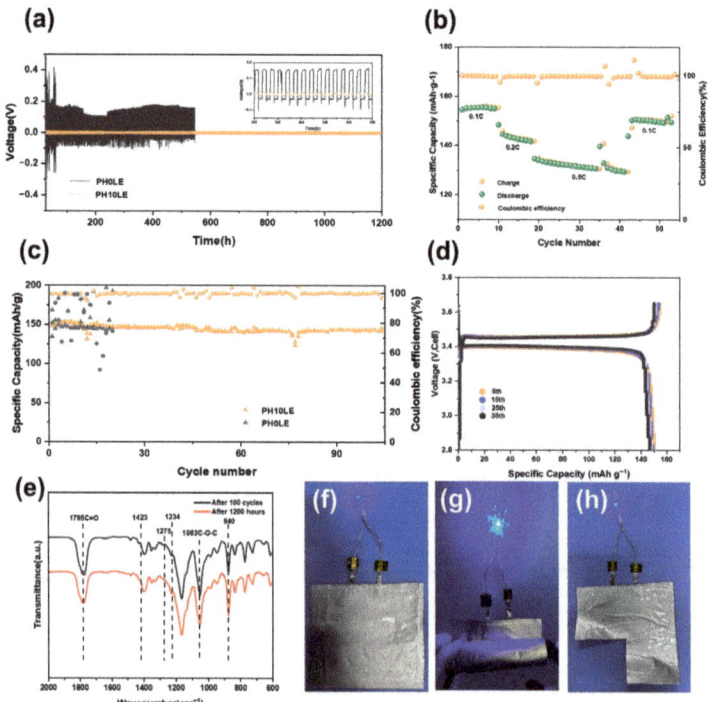

Figure 6. (**a**) Voltage profiles of Li | PH0LE | Li and Li | PH10LE | Li symmetrical cells at room temperature, and the inset in (**a**) shows magnified voltage profiles in an hour range of 340 h to 350 h. (**b**) Rate performance of Li/PHxLE/LFP (x = 0 or 10) coin cells at room temperature. (**c**) Cycling performance of Li/PHxLE/LFP (x = 0 or 10) coin cells at 0.1 C. (**d**) Charge–discharge curves of Li | PH10LE | LFP cell at different cycle numbers. FT-IR spectrum of (**e**) FCPE after 100 cycles of Li/FCPE/LFP and FCPE after 1200 h of cycling in Li/FCPE/Li. (**f**–**h**) Digital photograph of a blue LED light illuminated by an Li/FCPE/LFP pouch cell.

To validate the practical performance of this approach, a Li/FCPE/LFP pouch cell was assembled. After charging the pouch cell at a rate of 0.1 C to full capacity, no swelling phenomenon occurred. As shown in Figure 6e–h, at room temperature, the pouch cell was able to power a blue LED light normally. Furthermore, even when the pouch cell underwent arbitrary bending and shearing, it continued to supply power to the blue LED light without affecting its normal operation. This substantiates and underscores the extensive potential application of FCPE as elucidated in the study, indicating its promising utilization in flexible wearable electronic devices. In order to elucidate that no side reactions occur in the formation of the SEI (solid electrolyte interface) film during the battery cycling process in any component of the solid-state electrolyte, infrared tests were conducted on the FCPE of Li/FCPE/LFP after 100 cycles and Li/FCPE/Li after 1200 h of cycling (Figure 6e). The test outcomes reveal negligible deviation in characteristic peaks, signifying the ongoing stability of the solid electrolyte diaphragm structure throughout the cycling process. This further substantiates the cyclic stability of the solid electrolyte. The pouch cell assembled using PH10LE also exhibits excellent performance, as shown in Figure S3, achieving a capacity retention rate of 153.51 mA h g^{-1} at 0.1 C. This promising result suggests potential applications in future wearable electronic devices.

Based on the findings presented in the research, the augmentation mechanisms attributed to FCPE in enhancing lithium battery performance can be distilled into several key facets. First, ceramic particles enveloped within PVDF-HFP, culminating in the establishment of a robust PVDF-HFP/LLZTO interface and provide sufficient mechanical strength. The resulting three-dimensional network structure serves to intricately interconnect the lithium-ion transport interface. Second, within the 3D net-like structure, the ion conductor PVEC-based electrolyte permeates, adeptly gathering lithium ions along the pore walls. This not only amplifies ion conductivity but also benefits from the advantageous properties of VEC monomers, thereby fortifying Li$^+$ ion conductivity and antioxidation capabilities. Complemented by a passive layer founded on LiF, the entire system exhibits commendable stability, both in the context of Li-metal interactions and overall operational endurance [49].

4. Conclusions

In summary, a universal and straightforward method has been developed for preparing a netlike flexible composite electrolyte. The results indicate that the 3D active nanofillers in the PVDF-HFP-based CPE exhibit high thermal stability and outstanding Li$^+$ conductivity. After filling the network with a PVEC-based electrolyte, the ion conductivity further increases to 1.21×10^{-4} S cm^{-1} at 25 °C. Moreover, the flexible composite electrolyte membrane creates a 3D net-like structure with multiple rapid pathways for Li$^+$ that effectively control Li$^+$ flux, leading to uniform lithium deposition. Consequently, the symmetrical lithium cells exhibited remarkable stability when using the FCPE, while the assembled Li/FCPE/LFP battery showcased exceptional cycling performance. FT-IR spectroscopy analysis of the solid electrolyte membrane post-cycling reveals that its structure remains stable, further confirming the structural stability of FCPE. This work introduces a novel strategy for advancing the enhancement of flexible all-solid-state batteries.

Supplementary Materials: The following supporting information can be downloaded at: https://www.mdpi.com/article/10.3390/polym16050565/s1, Figure S1: The FT-IR spectrum of PVEC-based electrolyte, Figure S2: The TGA curve of PH0L and PH10L, Figure S3: Charge–discharge curves of Li|PH10LE|LFP pouch cell at 0.1 C, Table S1: Ionic conductivity in all-solid-state lithium-ion batteries have been reported.

Author Contributions: Conceptualization, X.M. and D.M.; methodology, X.M.; software, W.X.; validation, S.Y., H.Z. and Y.Z.; formal analysis, X.M.; investigation, X.M.; resources, D.M.; data curation, X.M., X.L. and D.M.; writing—original draft preparation, X.M.; writing—review and editing, X.L., Z.Y. and J.L.; visualization, X.M.; supervision, D.D.; project administration, D.D. All authors have read and agreed to the published version of the manuscript.

Funding: This research was funded by National Natural Science Foundation of China (No.: 51902130, 52072085), Science and Technology Program of University of Jinan (No.: XKY2103, XKY2105).

Institutional Review Board Statement: "Not applicable" for studies not involving humans or animals.

Data Availability Statement: Data are contained within the article or Supplementary Materials.

Conflicts of Interest: The authors declare no conflicts of interest.

References

1. Chen, S.; Nie, L.; Hu, X.; Zhang, Y.; Zhang, Y.; Yu, Y.; Liu, W. Ultrafast Sintering for Ceramic-Based All-Solid-State Lithium-Metal Batteries. *Adv. Mater.* **2022**, *34*, 2200430. [CrossRef]
2. Du, F.; Zhao, N.; Li, Y.; Chen, C.; Liu, Z.; Guo, X. All solid state lithium batteries based on lamellar garnet-type ceramic electrolytes. *J. Power Sources* **2015**, *300*, 24–28. [CrossRef]
3. Feng, W.; Lai, Z.; Dong, X.; Li, P.; Wang, Y.; Xia, Y. Garnet-Based All-Ceramic Lithium Battery Enabled by Li$_{(2.985)}$B$_{(0.005)}$OCl Solder. *iScience* **2020**, *23*, 101071. [CrossRef] [PubMed]
4. Feng, W.; Yang, P.; Dong, X.; Xia, Y. A Low Temperature Soldered All Ceramic Lithium Battery. *ACS Appl. Mater. Interfaces* **2022**, *14*, 1149–1156. [CrossRef] [PubMed]
5. Yan, Y.; Ju, J.; Yu, M.; Chen, S.; Cui, G. In-situ Polymerization Integrating 3D Ceramic Framework in All Solid-state Lithium Battery. *J. Inorg. Mater.* **2020**, *35*, 1357–1364. [CrossRef]

6. Han, F.; Yue, J.; Chen, C.; Zhao, N.; Fan, X.; Ma, Z.; Gao, T.; Wang, F.; Guo, X.; Wang, C. Interphase Engineering Enabled All-Ceramic Lithium Battery. *Joule* **2018**, *2*, 497–508. [CrossRef]
7. Hao, F.; Liang, Y.; Zhang, Y.; Chen, Z.; Zhang, J.; Ai, Q.; Guo, H.; Fan, Z.; Lou, J.; Yao, Y. High-Energy All-Solid-State Organic–Lithium Batteries Based on Ceramic Electrolytes. *ACS Energy Lett.* **2020**, *6*, 201–207. [CrossRef]
8. Jin, Y.; Liu, C.; Zong, X.; Li, D.; Fu, M.; Tan, S.; Xiong, Y.; Wei, J. Interface engineering of $Li_{1.3}Al_{0.3}Ti_{1.7}(PO_4)_3$ ceramic electrolyte via multifunctional interfacial layer for all-solid-state lithium batteries. *J. Power Sources* **2020**, *460*, 228125. [CrossRef]
9. Yu, X.; Manthiram, A. A Long Cycle Life, All-Solid-State Lithium Battery with a Ceramic–Polymer Composite Electrolyte. *ACS Appl. Energy Mater.* **2020**, *3*, 2916–2924. [CrossRef]
10. Huo, H.; Chen, Y.; Luo, J.; Yang, X.; Guo, X.; Sun, X. Rational Design of Hierarchical "Ceramic-in-Polymer" and "Polymer-in-Ceramic" Electrolytes for Dendrite-Free Solid-State Batteries. *Adv. Energy Mater.* **2019**, *9*, 1804004. [CrossRef]
11. Jiang, T.; He, P.; Wang, G.; Shen, Y.; Nan, C.W.; Fan, L.Z. Solvent-Free Synthesis of Thin, Flexible, Nonflammable Garnet-Based Composite Solid Electrolyte for All-Solid-State Lithium Batteries. *Adv. Energy Mater.* **2020**, *10*, 1903376. [CrossRef]
12. Kang, Q.; Zhuang, Z.; Liu, Y.; Liu, Z.; Li, Y.; Sun, B.; Pei, F.; Zhu, H.; Li, H.; Li, P.; et al. Engineering the Structural Uniformity of Gel Polymer Electrolytes via Pattern-guided Alignment for Durable, Safe Solid-state Lithium Metal Batteries. *Adv. Mater.* **2023**, *35*, 2303460. [CrossRef] [PubMed]
13. Lee, M.J.; Han, J.; Lee, K.; Lee, Y.J.; Kim, B.G.; Jung, K.-N.; Kim, B.J.; Lee, S.W. Elastomeric electrolytes for high-energy solid-state lithium batteries. *Nature* **2022**, *601*, 217–222. [CrossRef] [PubMed]
14. Nie, Y.; Yang, T.; Luo, D.; Liu, Y.; Ma, Q.; Yang, L.; Yao, Y.; Huang, R.; Li, Z.; Akinoglu, E.M.; et al. Tailoring Vertically Aligned Inorganic-Polymer Nanocomposites with Abundant Lewis Acid Sites for Ultra-Stable Solid-State Lithium Metal Batteries. *Adv. Energy Mater.* **2023**, *13*, 2204218. [CrossRef]
15. Ren, Z.; Li, J.; Gong, Y.; Shi, C.; Liang, J.; Li, Y.; He, C.; Zhang, Q.; Ren, X. Insight into the integration way of ceramic solid-state electrolyte fillers in the composite electrolyte for high performance solid-state lithium metal battery. *Energy Storage Mater.* **2022**, *51*, 130–138. [CrossRef]
16. Shan, X.; Zhao, S.; Ma, M.; Pan, Y.; Xiao, Z.; Li, B.; Sokolov, A.P.; Tian, M.; Yang, H.; Cao, P.-F. Single-Ion Conducting Polymeric Protective Interlayer for Stable Solid Lithium-Metal Batteries. *ACS Appl. Mater. Interfaces* **2022**, *14*, 56110–56119. [CrossRef]
17. Wan, J.; Xie, J.; Kong, X.; Liu, Z.; Liu, K.; Shi, F.; Pei, A.; Chen, H.; Chen, W.; Chen, J.; et al. Ultrathin, flexible, solid polymer composite electrolyte enabled with aligned nanoporous host for lithium batteries. *Nat. Nanotechnol.* **2019**, *14*, 705–711. [CrossRef]
18. Wang, Y.; Liu, T.; Liu, C.; Liu, G.; Yu, J.; Zou, Q. Solid-state lithium battery with garnet $Li_7La_3Zr_2O_{12}$ nanofibers composite polymer electrolytes. *Solid State Ion.* **2022**, *378*, 115897. [CrossRef]
19. Wang, Y.; Zanelotti, C.J.; Wang, X.; Kerr, R.; Jin, L.; Kan, W.H.; Dingemans, T.J.; Forsyth, M.; Madsen, L.A. Solid-state rigid-rod polymer composite electrolytes with nanocrystalline lithium ion pathways. *Nat. Mater.* **2021**, *20*, 1255–1263. [CrossRef]
20. Yang, H.; Zhang, B.; Jing, M.; Shen, X.; Wang, L.; Xu, H.; Yan, X.; He, X. In Situ Catalytic Polymerization of a Highly Homogeneous PDOL Composite Electrolyte for Long-Cycle High-Voltage Solid-State Lithium Batteries. *Adv. Energy Mater.* **2022**, *12*, 2201762. [CrossRef]
21. Yang, L.; Wang, Z.; Feng, Y.; Tan, R.; Zuo, Y.; Gao, R.; Zhao, Y.; Han, L.; Wang, Z.; Pan, F. Flexible Composite Solid Electrolyte Facilitating Highly Stable "Soft Contacting" Li-Electrolyte Interface for Solid State Lithium-Ion Batteries. *Adv. Energy Mater.* **2017**, *7*, 1701437. [CrossRef]
22. Zheng, Y.; Yao, Y.; Ou, J.; Li, M.; Luo, D.; Dou, H.; Li, Z.; Amine, K.; Yu, A.; Chen, Z. A review of composite solid-state electrolytes for lithium batteries: Fundamentals, key materials and advanced structures. *Chem. Soc. Rev.* **2020**, *49*, 8790–8839. [CrossRef] [PubMed]
23. Chen, K.; Shen, X.; Luo, L.; Chen, H.; Cao, R.; Feng, X.; Chen, W.; Fang, Y.; Cao, Y. Correlating the Solvating Power of Solvents with the Strength of Ion-Dipole Interaction in Electrolytes of Lithium-ion Batteries. *Angew. Chem. Int. Ed.* **2023**, *62*, e202312373. [CrossRef] [PubMed]
24. Su, Y.; Zhang, X.; Du, C.; Luo, Y.; Chen, J.; Yan, J.; Zhu, D.; Geng, L.; Liu, S.; Zhao, J.; et al. An All-Solid-State Battery Based on Sulfide and PEO Composite Electrolyte. *Small* **2022**, *18*, 2202069. [CrossRef] [PubMed]
25. Baek, S.-W.; Lee, J.-M.; Kim, T.Y.; Song, M.-S.; Park, Y. Garnet related lithium ion conductor processed by spark plasma sintering for all solid state batteries. *J. Power Sources* **2013**, *249*, 197–206. [CrossRef]
26. Chen, W.-P.; Duan, H.; Shi, J.-L.; Qian, Y.; Wan, J.; Zhang, X.-D.; Sheng, H.; Guan, B.; Wen, R.; Yin, Y.-X.; et al. Bridging Interparticle Li^+ Conduction in a Soft Ceramic Oxide Electrolyte. *J. Am. Chem. Soc.* **2021**, *143*, 5717–5726. [CrossRef]
27. Yu, G.; Wang, Y.; Li, K.; Sun, S.; Sun, S.; Chen, J.; Pan, L.; Sun, Z. Plasma optimized $Li_7La_3Zr_2O_{12}$ with vertically aligned ion diffusion pathways in composite polymer electrolyte for stable solid-state lithium metal batteries. *Chem. Eng. J.* **2021**, *430*, 132874. [CrossRef]
28. Guo, J.; Zheng, J.; Zhang, W.; Lu, Y. Recent Advances of Composite Solid-State Electrolytes for Lithium-Based Batteries. *Energy Fuels* **2021**, *35*, 11118–11140. [CrossRef]
29. Chen, F.; Yang, D.; Zha, W.; Zhu, B.; Zhang, Y.; Li, J.; Gu, Y.; Shen, Q.; Zhang, L.; Sadoway, D.R. Solid polymer electrolytes incorporating cubic $Li_7La_3Zr_2O_{12}$ for all-solid-state lithium rechargeable batteries. *Electrochim. Acta* **2017**, *258*, 1106–1114. [CrossRef]

30. Liang, Y.F.; Deng, S.J.; Xia, Y.; Wang, X.L.; Xia, X.H.; Wu, J.B.; Gu, C.D.; Tu, J.P. A superior composite gel polymer electrolyte of $Li_7La_3Zr_2O_{12}$- poly(vinylidene fluoride-hexafluoropropylene) (PVDF-HFP) for rechargeable solid-state lithium ion batteries. *Mater. Res. Bull.* **2018**, *102*, 412–417. [CrossRef]
31. Lin, Y.; Liu, K.; Xiong, C.; Wu, M.; Zhao, T. A composite solid electrolyte with an asymmetric ceramic framework for dendrite-free all-solid-state Li metal batteries. *J. Mater. Chem. A* **2021**, *9*, 9665–9674. [CrossRef]
32. Pearse, A.J.; Schmitt, T.E.; Fuller, E.J.; El-Gabaly, F.; Lin, C.-F.; Gerasopoulos, K.; Kozen, A.C.; Talin, A.A.; Rubloff, G.; Gregorczyk, K.E. Nanoscale Solid State Batteries Enabled by Thermal Atomic Layer Deposition of a Lithium Polyphosphazene Solid State Electrolyte. *Chem. Mater.* **2017**, *29*, 3740–3753. [CrossRef]
33. Thomas-Alyea, K.E. Design of Porous Solid Electrolytes for Rechargeable Metal Batteries. *J. Electrochem. Soc.* **2018**, *165*, A1523–A1528. [CrossRef]
34. Jiang, Z.; Xie, H.; Wang, S.; Song, X.; Yao, X.; Wang, H. Perovskite Membranes with Vertically Aligned Microchannels for All-Solid-State Lithium Batteries. *Adv. Energy Mater.* **2018**, *8*, 1801433. [CrossRef]
35. Hu, J.; He, P.; Zhang, B.; Wang, B.; Fan, L.-Z. Porous film host-derived 3D composite polymer electrolyte for high-voltage solid state Lithium batteries. *Energy Storage Mater.* **2020**, *26*, 283–289. [CrossRef]
36. Guo, W.; Shen, F.; Liu, J.; Zhang, Q.; Guo, H.; Yin, Y.; Gao, J.; Sun, Z.; Han, X.; Hu, Y. In-situ optical observation of Li growth in garnet-type solid state electrolyte. *Energy Storage Mater.* **2021**, *41*, 791–797. [CrossRef]
37. Xie, Z.; Wu, Z.; An, X.; Yue, X.; Xiaokaiti, P.; Yoshida, A.; Abudula, A.; Guan, G. A sandwich-type composite polymer electrolyte for all-solid-state lithium metal batteries with high areal capacity and cycling stability. *J. Membr. Sci.* **2020**, *596*, 117739. [CrossRef]
38. Gao, L.; Li, J.; Ju, J.; Cheng, B.; Kang, W.; Deng, N. Polyvinylidene fluoride nanofibers with embedded $Li_{6.4}La_3Zr_{1.4}Ta_{0.6}O_{12}$ fillers modified polymer electrolytes for high-capacity and long-life all-solid-state lithium metal batteries. *Compos. Sci. Technol.* **2020**, *200*, 108408. [CrossRef]
39. Huo, H.; Li, X.; Chen, Y.; Liang, J.; Deng, S.; Gao, X.; Doyle-Davis, K.; Li, R.; Guo, X.; Shen, Y.; et al. Bifunctional composite separator with a solid-state-battery strategy for dendrite-free lithium metal batteries. *Energy Storage Mater.* **2019**, *29*, 361–366. [CrossRef]
40. Li, X.; Cong, L.; Ma, S.; Shi, S.; Li, Y.; Li, S.; Chen, S.; Zheng, C.; Sun, L.; Liu, Y.; et al. Low Resistance and High Stable Solid–Liquid Electrolyte Interphases Enable High-Voltage Solid-State Lithium Metal Batteries. *Adv. Funct. Mater.* **2021**, *31*, 2010611. [CrossRef]
41. Kalnaus, S.; Dudney, N.J.; Westover, A.S.; Herbert, E.; Hackney, S. Solid-state batteries: The critical role of mechanics. *Science* **2023**, *381*, eabg5998. [CrossRef]
42. Liu, H.; Li, J.; Feng, W.; Han, G. Strippable and flexible solid electrolyte membrane by coupling $Li_{6.4}La_3Zr_{1.4}Ta_{0.6}O_{12}$ and insulating polyvinylidene fluoride for solid state lithium ion battery. *Ionics* **2021**, *27*, 3339–3346. [CrossRef]
43. Peng, L.; Lu, Z.; Zhong, L.; Jian, J.; Rong, Y.; Yang, R.; Xu, Y.; Jin, C. Enhanced ionic conductivity and interface compatibility of PVDF-LLZTO composite solid electrolytes by interfacial maleic acid modification. *J. Colloid Interface Sci.* **2022**, *613*, 368–375. [CrossRef]
44. Shen, F.; Guo, W.; Zeng, D.; Sun, Z.; Gao, J.; Li, J.; Zhao, B.; He, B.; Han, X. A Simple and Highly Efficient Method toward High-Density Garnet-Type LLZTO Solid-State Electrolyte. *ACS Appl. Mater. Interfaces* **2020**, *12*, 30313–30319. [CrossRef] [PubMed]
45. Song, X.; Zhang, T.; Huang, S.; Mi, J.; Zhang, Y.; Travas-Sejdic, J.; Turner, A.P.; Gao, W.; Cao, P. Constructing a PVDF-based composite solid-state electrolyte with high ionic conductivity $Li_{6.5}La_3Zr_{1.5}Ta_{0.1}Nb_{0.4}O_{12}$ for lithium metal battery. *J. Power Sources* **2023**, *564*, 232849. [CrossRef]
46. Xie, H.; Li, C.; Kan, W.H.; Avdeev, M.; Zhu, C.; Zhao, Z.; Chu, X.; Mu, D.; Wu, F. Consolidating the grain boundary of the garnet electrolyte LLZTO with Li_3BO_3 for high-performance $LiNi_{0.8}Co_{0.1}Mn_{0.1}O_2$/$LiFePO_4$ hybrid solid batteries. *J. Mater. Chem. A* **2019**, *7*, 20633–20639. [CrossRef]
47. Rangasamy, E.; Wolfenstine, J.; Sakamoto, J. The role of Al and Li concentration on the formation of cubic garnet solid electrolyte of nominal composition $Li_7La_3Zr_2O_{12}$. *Solid State Ion.* **2012**, *206*, 28–32. [CrossRef]
48. Xu, Y.; Wang, K.; Zhang, X.; Ma, Y.; Peng, Q.; Gong, Y.; Yi, S.; Guo, H.; Zhang, X.; Sun, X.; et al. Improved Li-Ion Conduction and (Electro)Chemical Stability at Garnet-Polymer Interface through Metal-Nitrogen Bonding. *Adv. Energy Mater.* **2023**, *13*, 2204377. [CrossRef]
49. Zhou, Z.; Sun, T.; Cui, J.; Shen, X.; Shi, C.; Cao, S.; Zhao, J. A homogenous solid polymer electrolyte prepared by facile spray drying method is used for room-temperature solid lithium metal batteries. *Nano Res.* **2021**, *16*, 5080–5086. [CrossRef]
50. Zhang, W.; Nie, J.; Li, F.; Wang, Z.L.; Sun, C. A durable and safe solid-state lithium battery with a hybrid electrolyte membrane. *Nano Energy* **2018**, *45*, 413–419. [CrossRef]
51. Zhang, X.; Liu, T.; Zhang, S.; Huang, X.; Xu, B.; Lin, Y.; Xu, B.; Li, L.; Nan, C.-W.; Shen, Y. Synergistic Coupling between $Li_{6.75}La_3Zr_{1.75}Ta_{0.25}O_{12}$ and Poly(vinylidene fluoride) Induces High Ionic Conductivity, Mechanical Strength, and Thermal Stability of Solid Composite Electrolytes. *J. Am. Chem. Soc.* **2017**, *139*, 13779–13785. [CrossRef] [PubMed]
52. Fang, Z.; Zhao, M.; Peng, Y.; Guan, S. Poly (vinylidene fluoride) binder reinforced poly (propylene carbonate)/3D garnet nanofiber composite polymer electrolyte toward dendrite-free lithium metal batteries. *Mater. Today Energy* **2022**, *24*, 100952. [CrossRef]

Disclaimer/Publisher's Note: The statements, opinions and data contained in all publications are solely those of the individual author(s) and contributor(s) and not of MDPI and/or the editor(s). MDPI and/or the editor(s) disclaim responsibility for any injury to people or property resulting from any ideas, methods, instructions or products referred to in the content.

Article

Thermoelectric Properties of One-Pot Hydrothermally Synthesized Solution-Processable PEDOT:PSS/MWCNT Composite Materials

Haibin Li [1], Shisheng Zhou [2,3,*], Shanxiang Han [2], Rubai Luo [2,4,*], Jingbo Hu [2,4], Bin Du [2,3], Kenan Yang [1], Yizhi Bao [2], Junjie Jia [2] and Xuemei Zhang [2]

1. School of Mechanical and Precision Instrument Engineering, Xi'an University of Technology, Xi'an 710048, China; lhbzwt@163.com (H.L.); 1200210002@stu.xaut.edu.cn (K.Y.)
2. Faculty of Printing, Packaging Engineering and Digital Media Technology, Xi'an University of Technology, Xi'an 710048, China; s.x.han@stu.xaut.edu.cn (S.H.); hujingboxaut@163.com (J.H.); dubin@xaut.edu.cn (B.D.); baoyizhi0424@163.com (Y.B.); 13429762525@163.com (J.J.); 17791496292@163.com (X.Z.)
3. Shaanxi Provincial Key Laboratory of Printing and Packaging Engineering, Xi'an University of Technology, Xi'an 710048, China
4. Shanxi Key Laboratory of Advanced Manufacturing Technology, North University of China, Taiyuan 038507, China
* Correspondence: zhoushisheng@xaut.edu.cn (S.Z.); luorubai@xut.edu.cn (R.L.)

Citation: Li, H.; Zhou, S.; Han, S.; Luo, R.; Hu, J.; Du, B.; Yang, K.; Bao, Y.; Jia, J.; Zhang, X. Thermoelectric Properties of One-Pot Hydrothermally Synthesized Solution-Processable PEDOT:PSS/MWCNT Composite Materials. *Polymers* **2023**, *15*, 3781. https://doi.org/10.3390/polym15183781

Academic Editor: Jeong In Han

Received: 18 August 2023
Revised: 13 September 2023
Accepted: 14 September 2023
Published: 15 September 2023

Copyright: © 2023 by the authors. Licensee MDPI, Basel, Switzerland. This article is an open access article distributed under the terms and conditions of the Creative Commons Attribution (CC BY) license (https://creativecommons.org/licenses/by/4.0/).

Abstract: The combination of organic and inorganic materials has been considered an effective solution for achieving ambient thermoelectric energy harvesting and has been developing rapidly. Here, PEDOT:PSS/MWCNT (PPM) composite hydrogels were synthesized using the self-assembled gelation process of poly(3,4-ethylenedioxythiophene)-poly(styrenesulfonate) (PEDOT:PSS) and the interaction between PEDOT:PSS and multi-walled carbon nanotubes (MWCNTs) without the addition of any surfactant. After immersion in dimethyl sulfoxide and freeze-drying, the hydrogel is easily dispersed in water and used as a direct ink writing (DIW) 3D printing ink. At room temperature, the PPM-20 printed film with 20 wt% MWCNT solids achieved a maximum power factor of 7.37 $\mu W\ m^{-1}\ K^{-2}$ and maintained stable thermoelectric properties during repeated bending cycles. On this basis, a thermoelectric generator (TEG) consisting of five legs was printed, which could be produced to generate an open circuit voltage of 6.4 mV and a maximum output power of 40.48 nW at a temperature gradient of 50 K, confirming its great potential for application in high-performance flexible organic/inorganic thermoelectric materials.

Keywords: PEDOT:PSS/MWCNTs; thermoelectric ink; DIW printing

1. Introduction

Thermoelectric generators (TEGs) realize the conversion between thermal energy and electrical energy through the Seebeck effect, especially flexible TEGs, which have broad application prospects in the fields of flexible electronics, medical monitoring, and the Internet of Things by utilizing their waste heat recovery capabilities [1–3]. As the core component of TEGs, the types of thermoelectric materials include inorganic semiconductors, organic carbon materials, and conductive polymers [4,5]. The thermoelectric properties of materials are usually evaluated by the dimensionless figure of merit (ZT), ZT = $S^2 \sigma T/k$, where S, σ, T, and k stand for the Seebeck coefficient or thermopower, electrical conductivity, absolute temperature, and thermal conductivity, respectively. Compared with conventional inorganic thermoelectric materials, conducting polymers (CPs) have attracted extensive attention in the field of flexible electronics due to their inherent low cost, flexibility, and excellent solution processability [6,7]. Meanwhile, with the large-scale application of additive manufacturing technology, the demand for organic CPs materials is gradually increasing. Among them, the conjugated conductive material PEDOT:PSS exhibits high thermoelectric properties, low thermal conductivity, and a highly adjustable molecular structure or composition, which is considered one of the best candidates for organic thermoelectric materials. In particular, PEDOT:PSS-based thermoelectric materials have proven their ability to more efficiently fabricate high-performance flexible TEGs through additive manufacturing techniques [8–10].

In previous studies, in order to improve the carrier transport property and enhance thermoelectric performance, poly(3,4-ethylenedioxythiophene)-poly(styrenesulfonate) (PEDOT:PSS) has always been doped or post-treated with the help of organic solvents, acid or base solutions, and ionic liquids [11–13]. However, the low thermoelectric performance of the doped PEDOT:PSS is a key drawback for its use in practical applications. Therefore, the incorporation of nanomaterials into PEDOT:PSS has been demonstrated as an effective method to realize enhanced thermoelectric performance, which not only maintains a high σ value but also improves the S value [14]. In particular, there is a strong π–π interaction between carbon-based nanomaterials and PEDOT:PSS, which could reduce conjugated defects and lower the carrier hopping barrier. PEDOT:PSS/Carbon nanotubes (CNTs) composites are considered a promising strategy to enhance the thermoelectric properties of composites, and PEDOT:PSS aqueous solution can reduce the physical entanglement of CNTs [15–18]. For instance, Wei et al. [19] fabricated SWCNTs/PEDOT:PSS composite thermoelectric films by vacuum filtration combined with post-treatment and designed a TEG with an S-shaped architecture. After ionic liquid treatment, the conductivity was increased to 1562 ± 170 S cm^{-1}, and the Seebeck coefficient remained constant at 21.9 µV K^{-1}. He et al. [20] reported a ternary stretchable and flexible thermoelectric film based on PEDOT:PSS/CNT/WPU with Seebeck coefficient and conductivity of 31 µV K^{-1} and 18 S cm^{-1} at room temperature, respectively. The double-network conductive bridge composed of PEDOT:PSS and CNTs acts as both a thermoelectric material and a temperature and strain sensing medium, and the assembled sensor successfully detected temperature changes and strain deformation under self-powered conditions. However, most of the previous studies were mainly focused on the direct mixing of PEDOT:PSS/CNTs and further preparation of TEG by thin film techniques such as suction filtration and drop casting. Meanwhile, in a previous study, highly conductive PEDOT:PSS hydrogel fibers were successfully obtained through a sulfuric acid-assisted gelation process [21]. This inspired us to explore whether CNTs can be used as a reinforcing agent to composite with PEDOT:PSS through a gel process. To the best of our knowledge, no one has systematically investigated the thermoelectric properties of inks converted from self-assembled PEDOT:PSS/CNTs composite gels.

Compared with simple thin film preparation methods, additive manufacturing has unique advantages such as simple operation, high flexibility, and implementation of complex structures, and this technology is widely selected for the manufacture of new flexible devices [22,23]. Among them, direct ink writing (DIW) technology has low requirements for ink and is widely used in flexible sensors, robots, and biomedical equipment [24]. In this study, we report a simple one-pot hydrothermal method that allows the self-assembly of PEDOT:PSS solution into a three-dimensional hydrogel at an acidic environment. During this process, the added MWCNTs were uniformly and tightly bridged onto the PEDOT:PSS framework. The resulting hydrogel could be easily dispersed in water after freeze-drying and used as extrusion-based 3D printing ink. Compared to previous studies on the direct composites of PEDOT:PSS and MWCNTs for the preparation of thermoelectric inks, we propose for the first time the successful incorporation of MWCNTs into the PEDOT:PSS matrix during the gel formation process, resulting in a high-performance thermoelectric material that can be processed via solution methods. During the gel self-assembly formation process, the PEDOT chain coil conformation was transformed and PSS was partially removed from the PEDOT:PSS complex, which significantly improved the electrical conductivity of the inks. The introduction of MWCNTs and post-treatment with organic solvents greatly enhance the thermoelectric properties of the resulting composite thermoelectric ink. Among them, the PEDOT:PSS/MWCNT (PPM-20) composite ink, containing 20 wt% MWCNTs, exhibited good electrical conductivity and Seebeck coefficient when drop cast to prepare thermoelectric thin films, and could also withstand significant bending. Additionally, the performance of TEG prepared using DIW process was also investigated in this paper.

2. Materials and Methods

2.1. Materials

The PEDOT:PSS aqueous solution (Clevios PH1000, PSS:PEDOT = 2.5:1) was obtained from Germany Heraeus Co., Ltd. MWCNTs (Hanau, Germany) (purity higher than 85%, diameter of 5–8 nm) were purchased from China Macklin Co., Ltd. (Shanghai, China). All other reagents including dimethyl sulfoxide (DMSO) and sulfuric acid (H_2SO_4) were of laboratory grade (Macklin Co., Ltd., Shanghai, China). All the materials were used without further purification.

2.2. Fabrication of PEDOT:PSS/MWCNT Composite Aerogels, Inks, and Thermoelectric Films

The different PEDOT:PSS/MWCNT composite thermoelectric inks were obtained by dispersing self-assembled PEDOT:PSS/MWCNT aerogels with different MWCNT contents. Among them, the prerequisites for preparing PEDOT:PSS/MWCNT aerogels is to synthesize the composite hydrogels using a one-pot hydrothermal method. Firstly, different masses of MWCNTs (0, 1.89, 4, 6.35, 9, 12, 15.4, or 24 mg) were dispersed in PEDOT:PSS dispersion (3 mL) using ultrasonication in an ice bath. After stirring at room temperature for 1 h, 0.05 M H_2SO_4 was added to the composite system and the mixtures were further stirred for 10 min. Then, the homogeneous mixtures were transferred to a Teflon liner and subjected to a hydrothermal reaction at 90 °C for 3 h. After cooling to room temperature, the resulting cylindrical composite hydrogels were naturally cooled and soaked in DMSO for 6 h. Subsequently, the treated hydrogels were repeatedly washed with deionized water (DI) and placed in the cold trap of a freeze-dryer at −65 °C for 6 h. After 48 h of vacuum freeze-drying, the PEDOT:PSS/MWCNT hydrogels with different MWCNT contents were successfully transformed into corresponding composite aerogels. Finally, the composite aerogels were dispersed in water with the assistance of ultrasonication to prepare thermoelectric inks with a solids content of 2 wt% and stored in a refrigerator at 4 °C.

To fabricate free-standing films, the ink was drop casted on PET substrates pre-treated with UV/O_3 for 5 min for better wetting with the aqueous solution. Then, the film was annealed at 130 °C for 30 min and cut to a size of 35 mm × 7 mm for measurement. The thickness of the films, measured by a profilometer (Alpha-Step, KLA Tencor, Milpitas, CA, USA), was in a range between 100 and 200 μm.

2.3. Fabrication of Flexible TEG

The flexible TE generators based on PEDOT:PSS/MWCNT composite ink were fabricated by the DIW process. The printing procedure was conducted based on an in-house developed 3D printer, and using Repetier Host application to process the thermoelectric array model convert it into G-code (Figure S1). Five thermoelectric legs were printed on a PET substrate pre-treated with UV/O_3 using ink dispersion (nozzle diameter: 0.5 mm, print speed: 300 mm/min). After the first printing, thermoelectric arrays were dried on a hotplate at 85 °C, and at 65 °C for the subsequent print layers. To achieve good uniformity and thickness, 7 layers were printed repeatedly. Finally, the thermoelectric legs were connected in series using conductive silver paste, and the whole TEG was dried in an oven at 120 °C for 1 h.

2.4. Measurements and Characterizations

The morphologies of the composite aerogels were analyzed by using scanning electron microscopy (SEM Hitachi S4700, Tokyo, Japan) at an acceleration voltage of 15 kV. Fourier transform infrared spectroscopy (FT-IR) was collected on an IR-spirit FTIR spectrometer (Shimadzu, Kyoto, Japan) using attenuated total reflection (ATR) mode from 4000 to

400 cm^{-1} with 20 scans. X-ray photoelectron spectroscopy (XPS) spectra was recorded on an AXISULTRA spectrometer (Kratos, Manchester, UK) with monochromatic Al Kα (1486.71 eV) line at a power of 100 W (10 mA, 10 kV). The measurement of electrical conductivity was carried out using four-probe measuring instrument (Four Probe Technology RST-9, Guangzhou, China). The Seebeck coefficients of all samples were measured using a custom-made system consisting of a digital source meter (Keithley-2460, Tektronix, Beaverton, OR, USA), a controlled Peltier heater, a type K thermocouple, and a power supply. The Seebeck coefficient was determined by linearly fitting ΔV/ΔT to ΔV values measured at ten different ΔT values, taking five data points at each temperature. The measured absolute Seebeck coefficients (20.2 ± 0.4 µV K^{-1} at 20 °C) of pure nickel foil were used for calibration. The output performance of the TEG was tested at a temperature difference of 50 K by connecting a variable resistance box in series, where current and voltage values were obtained from a series ammeter and a parallel voltmeter, respectively.

3. Results and Discussion

Figure 1a schematically illustrates the manufacturing process of PEDOT:PSS/MWCNT composite inks, which includes the self-assembly gelation, the immersion treatment, the freeze-drying, and the subsequent dispersion process. In brief, the PEDOT:PSS dispersions containing MWCNTs were used as precursors for hydrothermal reactions. With the promotion of H$_2$SO$_4$, stable composite hydrogels were obtained by using π–π interactions and van der Waals forces between MWCNTs and PEDOT:PSS. The aerogel obtained after further freeze-drying of the hydrogel could be easily dispersed into DI through sonication, and the dispersion was well suitable for the DIW process (Figure 1b). To distinguish them, the composites were referred to as PPM, and according to the mass ratio of MWCNTs (0, 5, 10, 15, 20, 25, 30, and 40 wt%) in in the composite system correspondingly they were denoted as PPM-0, PPM-5, PPM-10, PPM-15, PPM-20, PPM-25, PPM-30, and PPM-40. For comparison, the original PEDOT:PSS solution (PH1000) and the PEDOT:PSS gel (PPG) without DMSO treatment were also prepared as control samples. It is noteworthy that composite solutions with a mass ratio of MWCNTs greater than 50 wt% have difficulty in forming stable gels by means of the self-assembly process of PEDOT:PSS, indicating that excessive MWCNT agglomeration disrupts the gel network.

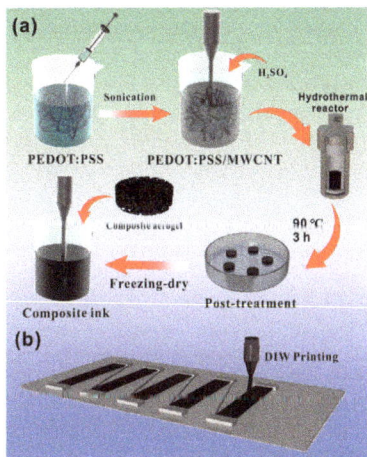

Figure 1. Schematic illustration of (**a**) the preparation of the PPM composite ink and (**b**) the printing process of TEG.

Figure 2 displays the morphology of PPM films with different MWCNT contents. It is obvious that the surface of the pure PPM-0 film is homogeneous and flat (Figure 2a). After the introduction of MWCNTs, the surface of the composite film gradually became rough until the interwoven MWCNTs could be clearly observed. When the MWCNT content is below 10 wt%, the few MWCNTs in the composite film were randomly distributed in the PEDOT:PSS matrix, and the less interconnected MWCNTs in PEDOT:PSS may affect the contact resistance between conductive PEDOTs. As the MWCNT content increases to 20 wt%, interwoven network-like structures of MWCNTs were clearly observed on the film surface, which is favorable for improving the Seebeck coefficient. In addition, the surface of MWCNTs is covered by PEDOT:PSS, which facilitates electron carrier mobility between MWCNTs and also results in low contrast of SEM images. However, when the content of MWCNTs was further increased to 40 wt%, some larger agglomerations appeared on the surface of the films, which might affect the electrical conductivity of the composite films.

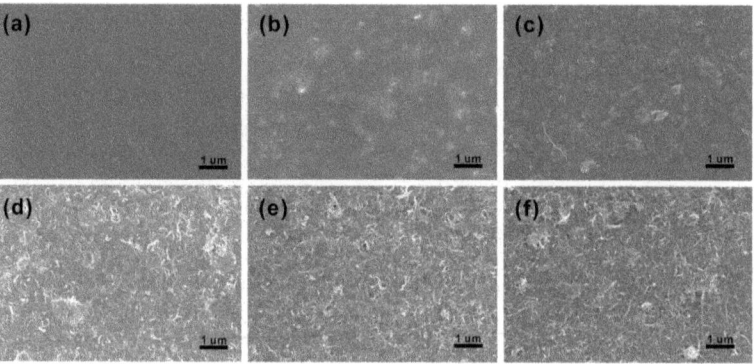

Figure 2. SEM surface images of (**a**) PPM-0, (**b**) PPM-5, (**c**) PPM-10, (**d**) PPM-20, (**e**) PPM-30, and (**f**) PPM-40 films.

Figure 3 shows the TE properties of the prepared PPG and PPM composite films with different MWCNT concentrations. As shown in Figure 3a, compared with the original PEDOT:PSS, the gel formation process, the introduction of MWCNTs and the treatment of the gel with DMSO immersion are effective in improving the electrical conductivity of the film. In particular, the electrical conductivity of PPM-0 films prepared using PEDOT:PSS hydrogels treated with DMSO was significantly increased to 218 S cm^{-1}. This apparent enhancement is attributed to the increased carrier mobility induced by the removal of PSS and the transformation of the PEDOT laminar structure. This is because in the original PEDOT:PSS, there is only one bond (C-C) between the carbon atoms of the EDOT monomer benzene ring, which corresponds to the coil-like benzene-fused structure. After the self-assembled gelation process and post-treatment with DMSO, the randomly coiled benzene-fused structure transforms into an extended quinoid structure (C=C) (Figure 3d) [25]. This transformation allows the formation of longer chains with stronger electron affinity and better conjugation properties, which can enhance the conductivity of PEDOT. Additionally, the formation of quinoid structure in PEDOT:PSS materials also contributes to improving the stability and durability of the material. Despite its excellent conductivity performance, the Seebeck coefficient of the post-treated PPM-0 film was significantly decreased. The conductivity of the PPM composite films decreased and the Seebeck coefficient increased after the addition of 5 wt% MWCNTs. This is due to the decrease in carrier concentration caused by the energy filtering effect between the n-type MWCNT and the p-type PEDOT:PSS interfaces, where the few and discontinuous MWCNTs act as resistive junctions in the highly conductive PEDOT:PSS. As illustrated in Figure 3d, carrier transport at low MWCNT concentrations occurs through the MWCNT-PEDOT:PSS–MWCNT junction. With the increase in the MWCNT content from 5 wt% to 20 wt%, the conductivity of the PPM composite film consistently increased to 185.19 S cm^{-1} due to a gradual reduction in the internal PEDOT:PSS–MWCNT junction, allowing for a more continuous conductive pathway. However, in the case of high MWCNT content, the conductivity tends to decrease because of the excessive MWCNT self-aggregation and excessive coverage of PEDOT:PSS, which is consistent with what was observed in the SEM images. Most notably, the Seebeck coefficient and electrical conductivity of the PPM composite films do not exhibit a monotonic relationship; that is, (i.e., the values of S and σ show an opposite trend). As shown in Figure 3b,c the addition of a small amount of MWCNTs can cause a significant increase in the Seebeck coefficient of the PPM composite film. The interfacial energy filtering effect between the two materials allows the passage of high-energy carriers and enhances the interfacial density [26]. The highest Seebeck coefficient of 22.8 μV K^{-1} has been achieved for the PPM-15 composite film, which is more than two times higher than that of PPM-0. As the concentration of MWCNTs increases, the n-type thermoelectric properties may intensify the phonon-polariton interaction, and the Seebeck coefficient of the composite film decreases slightly. The previous studies have indicated that the phonon-polaron interaction leads to scattering processes in electron transport, thereby affecting the magnitude of the Seebeck coefficient. Enough MWCNTs promote the propagation of phonons while also causing reverse scattering of low-energy and high-energy charge carriers, resulting in a shorter average free time for electron transport and a smaller Seebeck coefficient [27]. Combined with conductivity and Seebeck coefficient analysis, the PPM-20 composite film exhibited the highest power factor (PF = $S^2\sigma$) of 7.37 μW m^{-1} K^{-2}. According to previous studies, the power factor of the directly dispersed PEDOT:PSS/MWCNT composite was only 0.09 [28]. It is noteworthy that the PEDOT:PSS in all of the prepared PPM composites was not doped beforehand. This means that the PPM composite films obtained by gelation, immersion, and then cold-dry dispersion have great potential for TE applications, and more studies are still needed to further understand this point.

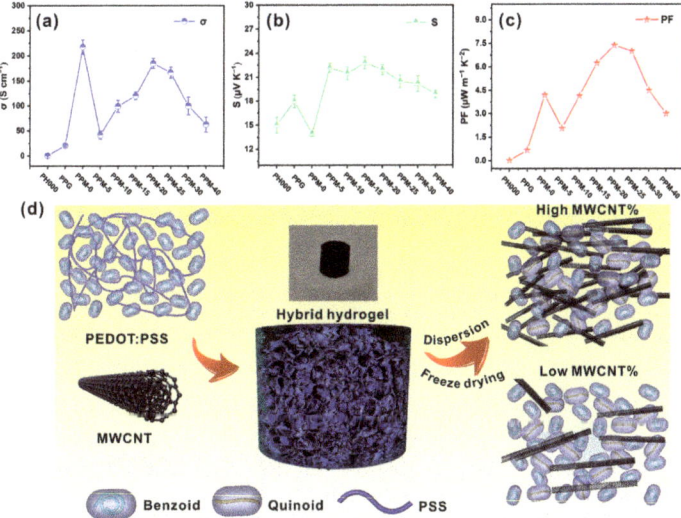

Figure 3. Room-temperature TE performances of all samples including (**a**) electrical conductivity, (**b**) Seebeck coefficient, and (**c**) power factor. (**d**) Conceptual illustration of the distribution of PEDOT:PSS and MWCNTs in different periods of composite morphology.

In order to elucidate the mechanism of the enhanced thermoelectric properties of PPM composite materials, Fourier infrared spectroscopy and XPS techniques have been employed. As shown in Figure 4a, in the spectrum of the CNT-free ink, the band at 1278 cm^{-1} is due to the C-C stretching vibration of the thiophene ring. The peaks at 1008 and 1125 cm^{-1} are assigned to the asymmetric vibration of S-O in PSS macromolecules and sulfonate groups, respectively. Meanwhile, the spectra of inks exhibit a strong broad band at 3300 cm^{-1}, which is caused by the O-H stretching modes of the hydroxyl groups, and this peak gradually increases with the superposition of the treatments [29]. Compared to PH1000, the peak of C=C stretching of quinoid structure at 1637 cm^{-1} red-shifts to 1632 cm^{-1} in PPG and PPM ink, which indicates the transformation of the benzoid structure of PEDOT:PSS to quinone structure during gelation [30]. In addition, from the FT-IR spectra of PPM-20 composite ink, some of the characteristic peaks of PEDOT:PSS are masked, indicating a strong interaction between PEDOT:PSS and MWCNTs.

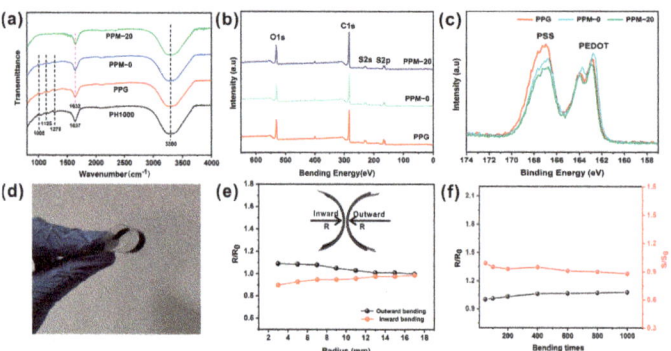

Figure 4. (**a**) FTIR spectra of pure PEDOT-PSS, PPG, PPM-0, and PPM-20. (**b**) XPS full spectra and (**c**) S$_{2p}$ spectra of PPG, PPM-0, and PPM-20. (**d**) Digital photos of flexible PPM-20 composite film. (**e**) The relative change in resistance for different inward or outward bending radiuses. (**f**) The relative change in conductivity and Seebeck coefficient at different bending cycles for a bending radius of 7 mm.

XPS analysis was used to investigate the change in the PSS component during the gelation process and subsequent solvent post-treatment process. Figure 4b,c show the full spectra and S$_{2p}$ XPS spectra of the PPG, PPM-0, and PPM-20 films, which reveal that the thermoelectric films are composed mainly of elements C, S, and O. In Figure 4c, the peak at 170–166 eV is associated with the sulfur atom of the sulfonic acid group in PSS chains [31]. While the two peaks at 162–166 eV are caused by the spin-orbit coupling of the C-S bond of the thiophene ring in the PEDOT chains, the peak areas are usually used to estimate the relative amounts of PEDOT and PSS [32]. Compared to the original PH1000 (2.0), the PSS and PEDOT ratios for PPG, PPM-0, and PPM-20 were 1.39, 1.26, and 1.21. These results demonstrate that the self-gelation process and post-treatment are effective in removing PSS, resulting in an increase in PEDOT-rich domains. This leads to a further enhancement of the thermoelectric properties. In particular, the screen effect of the DMSO bath can substantially reduce the Coulombic forces between PEDOT and PSS, which is consistent with the trend of thermoelectric properties in Figure 3. In addition, given the further decrease in the PSS/PEDOT ratio of PPM-20, the π–π interaction between MWCNTs and PEDOT:PSS also induces phase separation to some extent during the self-gelling process, resulting in an ordered lamellar structure that is more favorable for the transport of charge carriers.

As shown in Figure 4d, PPM-20 thermoelectric films exhibit favorable flexibility. Considering the stability required in the practical application of flexible thermoelectric components, we further investigated the TE performance of the PPM-20 composite film during bending (Figure 4e,f). The front and back of the film are mounted on separate cylinders surfaces of different radii so that they undergo different degrees of inward and outward bending. The results suggest that the film presents only a small increase or decrease in resistance, even at a bending radius of 3 mm. Furthermore, after as many as 1000 bends, the resistance and Seebeck coefficient of the PPM-20 film did not change significantly, displaying excellent stability and durability, which is beneficial for practical applications.

In order to test the power generation characteristics of the PPM-20 thermoelectric material, five thermoelectric legs were printed on the PET substrate using the DIW process. They were connected in series with silver paste to form a TEG (Figure 5a). Figure 5b shows a sketch of the circuit schematic assembled to measure the performance of the TEG, in which the load resistance is controlled by an adjustable resistor module. It can be seen from Figure 5c that the open-circuit voltage generated by the TEG increases almost linearly with the temperature gradient (ΔT = 0–50 K). In addition, the output voltage, output current, and output power of the TEG were measured with different external load resistors at a stable temperature difference of 50 K (Figure 5c). The output power of the TEG can be determined using the following equations:

$$P = \left(\frac{U}{R_{in} + R_{load}}\right)^2 R_{load} \qquad (1)$$

where R_{in} is internal resistance and R_{load} is load resistance. As the external load resistance varies, the output voltage and current are inversely proportional, and the output power exhibits a parabolic relationship with the current. TEG produces a maximum output power of 40.48 nW when the load resistance is equal to the internal resistance of the TEG. Moreover, as shown in Figure 5d, the beaker with the TEG attached was continuously poured with 70 °C hot water and reached the lower edge of the TEG, producing a potential difference of 5.6 mV. This indicates that the PPM composite film can effectively trap waste heat from the environment. For wider commercial applications, the performance of PPM-20 thermoelectric devices should be further improved. Doping modifications to PEDOT:PSS was able to substantially improve the electrical properties, which may have a direct impact on the thermoelectric properties of the PPM composite inks.

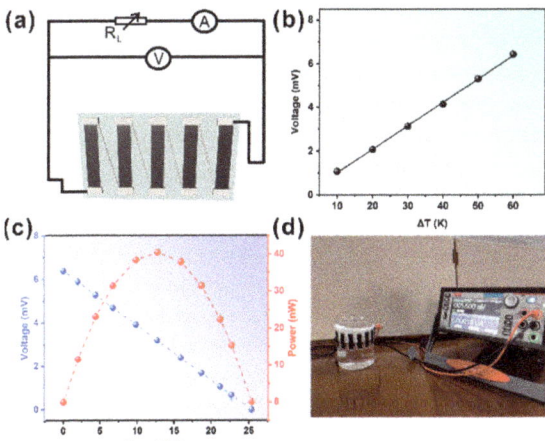

Figure 5. (a) Schematic diagram of the circuit for evaluating the output characteristics of the five-leg TEG. (b) The relationship between the output voltage and temperature gradient with five legs. (c) Voltage–current–power output curves at ΔT = 50 K. (d) Photograph of the 5.6 mV voltage created when the TEG was wrapped around a beaker and 70 °C warm water was poured in until it touched the underside of the TEG.

4. Conclusions

In summary, a simple one-pot hydrothermal method has been developed for the synthesis of solution-processable PPM composites. During the self-assembly of the gel, MWCNTs are firmly bridged to the PEDOT:PSS three-dimensional backbone due to π–π interactions and van der Waals forces between PEDOT:PSS and MWCNTs. The PPM aerogel obtained by freeze-drying is well dispersed in water and has been proven to be suitable for use in the DIW process. The energy filtering effect induced by the introduction of a small amount of MWCNTs in the composite system filters the low energy carriers and improved the Seebeck coefficient. The PPM-20 composite film showed a maximum PF value of 7.37 µW m at a solid content of MWCNTs of 20 wt%. In addition, the printed PPM-20 film has good flexibility and stable thermoelectric properties, which can withstand a wide range of repetitive bending variations. Finally, a TEG consisting of 5 legs manufactured by the fully printed process generated a maximum output power of 40.48 nW at a temperature difference of 50 K. This method of synthesizing solution-processable PEDOT:PSS/MWCNTs inks by self-assembly has great potential to facilitate the development and application of printable organic/inorganic composite thermoelectric materials.

Supplementary Materials: The following supporting information can be downloaded at: https://www.mdpi.com/article/10.3390/polym15183781/s1, Figure S1: Photos of a self-made 3D printing extrusion system and Repetier Host application interface.

Author Contributions: H.L. Conceptualization, Methodology, and Writing—original draft. S.Z. Resources and Supervision. S.H. Resources and Supervision. R.L. Formal analysis and Investigation. S.H. Funding acquisition and Visualization. J.H. Methodology and Writing—review & editing. B.D. Software and Validation. K.Y. Resources and Validation. Y.B. Project administration. J.J. Formal analysis. X.Z. Supervision. All authors have read and agreed to the published version of the manuscript.

Funding: This work is supported by the Xi'an Science and technology plan project under Grant Nos. GX2338, the Key scientific research program of Shaanxi Provincial Department of Education under Grant Nos. 22JY046 and Nos. 21JY032, the Opening Project of Shanxi Key Laboratory of Advanced Manufacturing Technology of North University of China under Grant Nos. XJZZ202104, and the General project of natural science basic research program of Shaanxi Provincial Department of Science and Technology under Grant Nos. 2023-JC-YB-424. All authors have read and agreed to the published version of the manuscript.

Institutional Review Board Statement: Not applicable.

Data Availability Statement: The datasets used and analyzed in the current study are available from the corresponding author on reasonable request.

Conflicts of Interest: There are no conflict of interests to declare.

References

1. Shi, W.; Wang, D.; Shuai, Z. High-performance organic thermoelectric Materials: Theoretical insights and computational design. *Adv. Electron. Mater.* **2019**, *5*, 1800882. [CrossRef]
2. Chen, Z.G.; Shi, X.; Zhao, L.D.; Zou, J. High-performance SnSe thermoelectric materials: Progress and future challenge. *Prog. Mater. Sci.* **2018**, *97*, 283–346. [CrossRef]
3. Zhang, Z.; Chen, G.; Wang, H.; Li, X. Template-directed in situ polymerization preparation of nanocomposites of PEDOT:PSS-coated multi-walled carbon nanotubes with enhanced thermoelectric property. *Chem. Asian J.* **2015**, *10*, 149–153. [CrossRef]
4. Panigrahy, S.; Kandasubramanian, B. Polymeric thermoelectric PEDOT: PSS & composites: Synthesis, progress, and applications. *Eur. Polym. J.* **2020**, *132*, 109726. [CrossRef]
5. Yun, J.S.; Choi, S.; Im, S.H. Advances in carbon-based thermoelectric materials for high-performance, flexible thermoelectric devices. *Carbon Energy* **2021**, *3*, 667–708. [CrossRef]
6. Li, Y.; Zhou, X.; Sarkar, B.; Gagnon-Lafrenais, N.; Cicoira, F. Recent progress on self-healable conducting polymers. *Adv. Mater.* **2022**, *34*, 2108932. [CrossRef] [PubMed]
7. Hao, Y.; He, X.; Wang, L.; Qin, X.; Chen, G.; Yu, J. Stretchable thermoelectrics: Strategies, performances, and applications. *Adv. Funct. Mater.* **2021**, *32*, 2109790. [CrossRef]
8. Kim, G.H.; Shao, L.; Zhang, K.; Pipe, K.P. Engineered doping of organic semiconductors for enhanced thermoelectric efficiency. *Nat. Mater.* **2013**, *12*, 719–723. [CrossRef]
9. Kee, S.; Haque, M.A.; Corzo, D.; Alshareef, H.N.; Baran, D. Self-healing and stretchable 3D-printed organic thermoelectrics. *Adv. Funct. Mater.* **2019**, *29*, 1905426. [CrossRef]
10. Shakeel, M.; Rehman, K.; Ahmad, S.; Amin, M.; Iqbal, N.; Khan, A. A low-cost printed organic thermoelectric generator for low-temperature energy harvesting. *Renew. Energy* **2021**, *167*, 853–860. [CrossRef]
11. Hosseini, E.; Kollath, V.O.; Karan, K. The key mechanism of conductivity in PEDOT:PSS thin films exposed by anomalous conduction behaviour upon solvent-doping and sulfuric acid post-treatment. *J. Mater. Chem. C* **2020**, *8*, 3982. [CrossRef]

12. Modarresi, M.; Zozoulenko, I. Why does solvent treatment increase the conductivity of PEDOT:PSS? Insight from molecular dynamics simulations. *Phys. Chem. Chem. Phys.* **2022**, *24*, 22073. [CrossRef] [PubMed]
13. Deng, W.; Deng, L.; Li, Z.; Zhang, Y.; Chen, G. Synergistically boosting thermoelectric performance of PEDOT:PSS/SWCNT composites via the ion-exchange effect and promoting SWCNT dispersion by the ionic liquid. *ACS Appl. Mater. Interfaces* **2021**, *13*, 12131–12140. [CrossRef] [PubMed]
14. Yang, J.; Jia, Y.; Liu, Y.; Liu, P.; Wang, Y.; Li, M.; Jiang, F.; Lan, X.; Xu, J. PEDOT:PSS/PVA/Te ternary composite fibers toward flexible thermoelectric generator. *Compos. Commun.* **2021**, *27*, 100855. [CrossRef]
15. Kim, D.; Kim, Y.; Choi, K.; Grunlan, J.C.; Yu, C. Improved thermoelectric behavior of nanotube-filled polymer composites with poly(3,4-ethylenedioxythiophene) poly (styrenesulfonate). *ACS Nano* **2010**, *4*, 513–523. [CrossRef] [PubMed]
16. Liu, S.; Li, H.; He, C. Simultaneous enhancement of electrical conductivity and seebeck coefficient in organic thermoelectric SWNT/PEDOT:PSS nanocomposites. *Carbon* **2019**, *149*, 25–32. [CrossRef]
17. Tonga, M.; Wei, L.; Lahti, P.M. Enhanced thermoelectric properties of PEDOT:PSS composites by functionalized single wall carbon nanotubes. *Int. J. Energy Res.* **2020**, *44*, 9149–9156. [CrossRef]
18. Wang, Y.; Wu, S.; Yin, Q.; Jiang, B.; Mo, S. Poly (3,4-ethylenedioxythiophene)/polypyrrole/carbon nanoparticle ternary nanocomposite films with enhanced thermoelectric properties. *Polymer* **2021**, *212*, 123131. [CrossRef]
19. Wei, S.; Liu, L.; Huang, X.; Zhang, Y.; Liu, F.; Deng, L.; Bilotti, E.; Chen, G. Flexible and foldable films of SWCNT thermoelectric composites and an S-shape thermoelectric generator with a vertical temperature gradient. *ACS Appl. Mater. Interfaces* **2022**, *14*, 5973–5982. [CrossRef]
20. He, X.; Hao, Y.; He, M.; Qin, X.; Wang, L.; Yu, J. Stretchable thermoelectric-based self-powered dual-parameter sensors with decoupled temperature and strain sensing. *ACS Appl. Mater. Interfaces* **2021**, *13*, 60498–60507. [CrossRef]
21. Yao, B.; Wang, H.; Zhou, Q.; Wu, M.; Zhang, M.; Li, C.; Shi, G. Ultrahigh-conductivity polymer hydrogels with arbitrary structures. *Adv. Mater.* **2017**, *29*, 1700974. [CrossRef] [PubMed]
22. Tan, L.J.; Zhu, W.; Zhou, K. Recent progress on polymer materials for additive manufacturing. *Adv. Funct. Mater.* **2020**, *30*, 2003062. [CrossRef]
23. Du, Y.; Chen, J.; Meng, Q.; Dou, Y.; Xu, J.; Shen, S.Z. Thermoelectric materials and devices fabricated by additive manufacturing. *Vacuum* **2020**, *178*, 109384. [CrossRef]
24. Schwartz, J.; Boydston, A.J. Multimaterial actinic spatial control 3D and 4D printing. *Nat. Commun.* **2019**, *10*, 791. [CrossRef] [PubMed]
25. EL-Shamy, A. Acido-treatment of PEDOT:PSS/Carbon Dots (C_{Dots}) nano-composite films for high thermoelectric power factor performance and generator. *Mater. Chem. Phys.* **2021**, *257*, 123762. [CrossRef]
26. Wei, S.; Zhang, Y.; Lv, H.; Deng, L.; Chen, G. SWCNT network evolution of PEDOT:PSS/SWCNT composites for thermoelectric application. *Chem. Eng. J.* **2022**, *428*, 131137. [CrossRef]
27. Wang, Y.; Wu, S.; Zhang, R.; Du, K.; Yin, Q.; Jiang, B.; Yin, Q.; Zhang, K. Effects of carbon nanomaterials hybridization of Poly(3,4-ethylenedioxythiophene): Poly (styrene sulfonate) on thermoelectric performance. *Nanotechnology* **2021**, *32*, 445705. [CrossRef] [PubMed]
28. Liu, Y.; Liu, H.; Wang, J.; Zhang, X. Thermoelectric behavior of PEDOT:PSS/CNT/ graphene composites. *J. Polym. Eng.* **2018**, *38*, 381–389. [CrossRef]
29. Khasim, S.; Pasha, A.; Badi, N.; Lakshmi, M.; Mishra, Y.K. High performance flexible supercapacitors based on secondary doped PEDOT-PSS-graphene nanocomposite films for large area solid state devices. *RSC. Adv.* **2020**, *10*, 10526. [CrossRef]
30. Sun, C.; Li, X.; Zhao, J.; Cai, Z.; Ge, F. A freestanding polypyrrole hybrid electrode supported by conducting silk fabric coated with PEDOT:PSS and MWCNTs for high-performance supercapacitor. *Electrochim. Acta* **2019**, *317*, 42–51. [CrossRef]
31. Cao, X.; Zhang, M.; Yang, Y.; Deng, H.; Fu, Q. Thermoelectric PEDOT:PSS Sheet/SWCNTs composites films with layered structure. *Compos. Commun.* **2021**, *27*, 100869. [CrossRef]
32. Jeong, W.; Gwon, G.; Ha, J.H.; Kim, D.; Eom, K.J.; Park, J.H.; Kang, S.J.; Kwak, B.; Hong, J.I.; Lee, S.; et al. Enhancing the conductivity of PEDOT:PSS films for biomedical applications via hydrothermal treatment. *Biosens. Bioelectron.* **2021**, *171*, 112717. [CrossRef]

Disclaimer/Publisher's Note: The statements, opinions and data contained in all publications are solely those of the individual author(s) and contributor(s) and not of MDPI and/or the editor(s). MDPI and/or the editor(s) disclaim responsibility for any injury to people or property resulting from any ideas, methods, instructions or products referred to in the content.

Article

Preparation of Zinc Oxide with Core–Shell Structure and Its Application in Rubber Products

Zhibin Wang, Zhanfeng Hou, Xianzhen Liu, Zhaolei Gu, Hui Li and Qi Chen *

Key Laboratory of Rubber-Plastics, Ministry of Education/Shandong Provincial Key Laboratory of Rubber-Plastics, School of Polymer Science and Engineering, Qingdao University of Science and Technology, Qingdao 266042, China; 18742017071@163.com (Z.W.); 13012545139@163.com (Z.G.)
* Correspondence: 03293@qust.edu.cn

Abstract: Zinc oxide is a crucial component in rubber products, but its excessive usage can lead to environmental damage. As a result, reducing the amount of zinc oxide in products has become a critical issue that many researchers aim to address. This study employs a wet precipitation method to prepare ZnO particles with different nucleoplasmic materials, resulting in ZnO with a core–shell structure. The prepared ZnO underwent XRD, SEM, and TEM analysis, indicating that some of the ZnO particles were loaded onto the nucleosomal materials. Specifically, ZnO with a silica core–shell structure demonstrated 11.9% higher tensile strength, 17.2% higher elongation at break, and 6.9% higher tear strength compared to the indirect method of ZnO preparation. The core–shell structure of ZnO also helps reduce its application in rubber products, thereby achieving the dual objective of protecting the environment and improving the economic efficiency of rubber products.

Keywords: core–shell structured zinc oxide; natural rubber; tread rubber; reduction in zinc oxide dosage

Citation: Wang, Z.; Hou, Z.; Liu, X.; Gu, Z.; Li, H.; Chen, Q. Preparation of Zinc Oxide with Core–Shell Structure and Its Application in Rubber Products. *Polymers* **2023**, *15*, 2353. https://doi.org/10.3390/polym15102353

Academic Editor: Fahmi Zaïri

Received: 16 April 2023
Revised: 13 May 2023
Accepted: 16 May 2023
Published: 18 May 2023

Copyright: © 2023 by the authors. Licensee MDPI, Basel, Switzerland. This article is an open access article distributed under the terms and conditions of the Creative Commons Attribution (CC BY) license (https://creativecommons.org/licenses/by/4.0/).

1. Introduction

Zinc oxide is an inorganic filler widely used in the rubber industry, and it has several functions. First, zinc oxide acts as an active agent in rubber to promote the vulcanization reaction, which helps to accelerate the rate of vulcanization and increase the degree of vulcanization, thus improving the physical and mechanical properties of rubber. Second, ZnO is used as a filler of rubber composites intended for products exhibiting increased heat conductivity. Furthermore, zinc oxide acts as an antioxidant that absorbs harmful ultraviolet rays, oxygen, and ozone to avoid their oxidation and the aging of rubber, thus prolonging the service life of rubber [1–3]. In addition, zinc oxide can be used as an antimicrobial agent to kill microorganisms by releasing oxygen and zinc ions, thus increasing the durability and service life of rubber products. In summary, zinc oxide is a very important rubber filler in rubber because of its multiple roles in promoting vulcanization, antioxidation, and antibacterial activity [4–6].

The rubber tire industry represents the largest consumer of zinc oxide, with approximately 50% of total zinc oxide usage being attributed to this industry [7]. Currently, micron-sized ZnO is used primarily as the curing active agent in the tire industry, and only a portion of the ZnO is involved in the activation of the curing reaction. As a result, a significant amount of ZnO remains in the rubber in the form of micron-sized particles, and residual Zn is released into the environment during tire operation, contributing to environmental pollution [8,9]. Despite the important role of ZnO in sulfur vulcanization, its concentration in rubber compounds, especially those used in aquatic environments, must be reduced to, at least, below 2.5 wt%, because zinc oxide is classified as being toxic to aquatic life. According to European Union Regulation (EC) No. 1272/2008 on classification, labeling, and packaging of substances and mixtures, ZnO was classified as Aquatic Acute 1 with hazard statement H400: Very toxic to aquatic life and Aquatic Chronic 1 with hazard statement H410: Very toxic to aquatic life with long lasting effects. The precaution recommended in this regulation is defined as P273: Avoid release to the environment. The release of zinc from rubber products occurs during their manufacture, use (dust created during the abrasion of tires on road surfaces), and recycling or disposal in landfills. A potential source of zinc in groundwater can also be rubber granulates made from end-of-life tires used to build artificial sports fields. Taking this into account, the problem of reducing the amount of zinc in rubber products is essential [10].

With advances in synthesis technology, various methods have been employed to prepare ZnO nanoparticles as a substitute for conventional ZnO. The methods and conditions for preparing ZnO nanoparticles have been extensively investigated and can be broadly categorized as solid-phase, liquid-phase, and gas-phase methods based on the phase state of the reactants [11–14]. It has been demonstrated that decreasing the particle size of ZnO nanoparticles results in a rougher surface and uneven atomic steps, leading to increased contact surface and chemical activity [15–18]. In contrast, the particle size of ZnO nanoparticles is considerably smaller than that of ZnO, making it theoretically feasible to replace ZnO and reduce the amount of zinc used [19–21]. The use of high-specific-surface-area ZnO nanoparticles can increase the contact area between ZnO and rubber, improve the efficiency of ZnO in the activation reaction (by increasing the reaction rate and reducing energy consumption during vulcanization), and simultaneously reduce the amount of ZnO used without compromising the enhancement effect [22–25]. According to thermodynamic theory, smaller particle sizes result in reduced dispersion effectiveness. ZnO nanoparticles tend to agglomerate due to their small size and high specific surface energy, thus limiting their nano-effect [26–28]. To address this issue, researchers have employed strategies such as loading ZnO nanoparticles or preparing core–shell structured particles. For instance, Magdalena G. et al. [6] used a gel method to coat ZnO nanoparticles onto the surface of SiO_2 to investigate the effect of SiO_2@ZnO core–shell structured nanoparticles on the kinetics of carboxylated nitrile rubber. Yalan L. et al. [29] used a wet blending method to load ZnO onto the surface of cellulose fibers to study the dispersion of cellulose fibers in the rubber matrix and its impact on the mechanical strength of natural rubber. Furthermore, Zeinab D.G. et al. [30] prepared CoO.CaO/ZnO core–shell structured particles and examined their effect on the mechanical properties of nitrile butadiene rubber (NBR), with ZnO as the core and CoO and CaO as the shell. This approach improved the tensile strength of NBR and enhanced the compatibility between ZnO and NBR. In addition, some researchers have employed nanoscale active ZnO with a micron-level carrier coating structure as the starting material and added various low-molecular-weight PIBs as dispersion aids to the nano-active ZnO powder to achieve surface modification, thereby improving several surface properties such as agglomeration adsorption and dispersion. This strategy ultimately enhances the compatibility of nano-active ZnO with rubber materials.

In this study, core–shell structured zinc oxide nanoparticles were synthesized via the wet precipitation method, using various materials including calcium carbonate, barium sulfate, silicon dioxide, thermally cracked carbon black, and graphene as the core material. The resulting core–shell structured zinc oxide nanoparticles were characterized in terms of morphology, particle size, and the extent of zinc oxide loading onto the core material. Subsequently, the prepared zinc oxide nanoparticles were incorporated into the formulation of semi-steel radial tire tread rubber, and the effects on the vulcanization characteristics, mechanical properties, friction properties, aging properties, and dynamic thermomechanical properties of the rubber were studied.

2. Experimental Part
2.1. Experimental Materials

Zinc chloride used in the preparation was sourced from Shandong Xuanhai Chemical Co., Ltd. (Heze, China), whereas sodium carbonate was obtained from Tianjin Jinhui Pharmaceutical Group Co., Ltd. (Tianjin, China) The calcium carbonate used in the experiment was provided by Changzhou Calcium Carbonate Co., Ltd. (Changzhou, China)

For the preparation of core–shell structure zinc oxide of barium sulfate, Shandong Qiyi Chemical Technology Co., Ltd. (Weifang, China) provided the barium sulfate, and Shandong Bluestar Dongda Chemical Co., Ltd. (Zibo, China) provided the silicon dioxide for the core–shell structure zinc oxide of silica. The graphene used in the experiment was sourced from Chinese Academy of Sciences Chengdu Organic Chemistry Co., Ltd. (Chengdu, China), whereas the thermal cracking carbon black was obtained from Jiangxi Black Cat Carbon Black Co., Ltd. (Jingdezhen, China). The indirect method of zinc oxide used in the experiment was provided by Anqiu Hengshan Zinc Industry Co., Ltd. (Anqiu, China), whereas the other complexes were provided by Guangzhou Chemical Reagent Factory. To prepare the zinc oxide with core–shell structure, the aforementioned raw materials were used, and the experiment was carried out according to the previously described method.

2.2. Preparation of Zinc Oxide with Core–Shell Structures

To prepare core–shell structured ZnO particles, analytical grade zinc chloride, sodium, carbonate calcium carbonate ($CaCO_3$), barium sulfate ($BaSO_4$), silicon dioxide (SiO_2), pyrolysis of carbon black (CBp), grapheme oxide (GO), and distilled water were used. To prepare ZnO@$CaCO_3$, 138 g of zinc chloride was dissolved in 330 mL of distilled water. Alternatively, 108 g of sodium carbonate was added to 500 mL of distilled water. The prepared zinc chloride solution was added to the beaker with 81 g of calcium carbonate solids and stirred at 80 °C to make the zinc chloride solution infiltrate the surface of the calcium carbonate solids. Then, the prepared sodium carbonate solution was slowly added to beaker and the solution was continuously stirred at 80 °C in an oil bath. A viscous and honey-like gel was obtained after continuously stirring the solution for 30 min at 80 °C. The solution was washed with water three times and dried for 12 h. Lastly, for the calcination process, a small amount of provided powers were put in an alumina crucible before being placed into a 600 °C furnace. The heating rate was 5 °C/min, and the heating operation was 3 h [31]. Figure 1 presents the process of core–shell structured ZnO. The core–shell structured zinc oxide/barium sulfate, zinc oxide/silica, zinc oxide/pyrolysis of carbon black, and zinc oxide/graphene were prepared by replacing the core–shell materials in the same way.

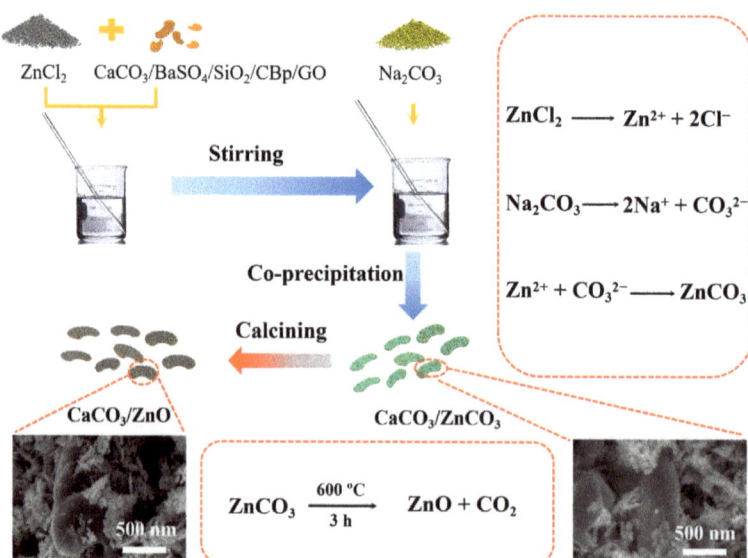

Figure 1. Preparation process of core–shell structure zinc oxide.

2.3. Preparation of Natural Rubber-Based Nanocomposites

The rubber and auxiliary were mixed in the compactor according to the formulation shown in Table 1. The total mixing time is 8 min and then vulcanized at 143 °C to obtain the best cure time (t_{90}). The rubber compounds obtained by adding different types of zinc oxide were named R-ZnO, R-ZnO@CaCO$_3$, R-ZnO@BaSO$_4$, R-ZnO@SiO$_2$, R-ZnO@CBp, and R-ZnO@GO.

Table 1. Tire tread rubber experimental formula (phr).

	R-ZnO	R-ZnO@Ca	R-ZnO@Ba	R-ZnO@Si	R-ZnO@CBp	R-ZnO@GO
BR9000	20	20	20	20	20	20
SBR	89	89	89	89	89	89
NR	15	15	15	15	15	15
Silica	65	65	65	65	65	65
N375	20	20	20	20	20	20
Indirect method ZnO	3	-	-	-	-	-
ZnO@CaCO$_3$	-	3	-	-	-	-
ZnO@BaSO$_4$	-	-	3	-	-	-
ZnO@SiO$_2$	-	-	-	3	-	-
ZnO@CBp	-	-	-	-	3	-
ZnO@Go	-	-	-	-	-	3
Si69	9.5	9.5	9.5	9.5	9.5	9.5
SA	2	2	2	2	2	2
S	1.3	1.3	1.3	1.3	1.3	1.3
CZ	1.8	1.8	1.8	1.8	1.8	1.8

2.4. Characterization

The crystal structure of the core–shell structured powdered ZnO was determined using a D-MAX2500/PC X-ray diffractometer (Nippon Rigaku Co., Ltd., Tokyo, Japan) with a test range of 10° to 80° and a scanning rate of 5°/min. The morphological characteristics were observed using a scanning electron microscope (JSM-7500F; Nippon Electron Co., Ltd., Tokyo, Japan) with an acceleration voltage of 3 kV. The transmission electron microscope (JEOL-JEM-2100; Japan Electron Co., Ltd., Tokyo, Japan) was used to investigate the microscopic morphology of ZnO with a core–shell structure made from different nucleosomal materials. To test the properties of the prepared core–shell structured ZnO, it was added to a natural rubber formulation. All vulcanization properties were measured using a vulcanometer (MDR2000, Alpha Technologies, Hudson, Ohio, USA) at 143 °C. The mechanical properties were tested using an electronic tensile machine (I-7000S, High Iron Co., Ltd., Taipei, Taiwan) with the sample strips cut into dumbbell-shaped strips with a length of 75 mm, thickness of 2.00 ± 0.03 mm, and working width of 4 mm. The tensile properties were measured at a speed of 500 mm/min. Tear strength was measured using the right-angle tear mode C according to ASTM D624. The thermal oxygen aging chamber was used to age the cut tensile and tear sample strips at 100 °C for 72 h. Abrasion was tested using a DIN abrasion machine (GT-7012-D, GOTECH Co., Ltd., Taichung, Taiwan) with a pressure of 10 N and a roller speed of 40 r/min. The dynamic mechanical analysis was performed using a dynamic thermomechanical analyzer (242, NETZSCH, Selb, Germany) with a test temperature range from −60 °C to 80 °C at a frequency of 3 Hz and a ramp rate of 3 °C/min.

3. Results and Discussion

3.1. Characterization of Core–Shell Structure ZnO

3.1.1. XRD

The crystallinity patterns and structures of the synthesized samples were analyzed using X-ray diffraction (XRD). Figure 2 shows the XRD patterns of commercially available ZnO composites prepared by the indirect method, which exhibited diffraction peaks at

31.6°, 34.4°, 36.1°, 47.3°, 56.3°, 62.6°, and 67.6° corresponding to (100), (002), (101), (102), (110), (103), and (112) planes. The core–shell structured ZnO prepared with nucleosomal materials such as graphene and pyrolysis of carbon black showed similar peak patterns, indicating that the structure of ZnO on the surface of CBp and GO is primarily in the hexagonal phase, which was produced at 600 °C. Hence, for the core–shell structured ZnO prepared with nucleosomal materials such as SiO_2, $CaCO_3$, and $BaSO_4$, some different diffraction peaks were detected in the patterns. The XRD patterns of Ca and Ba exhibited characteristic diffraction peaks of calcium carbonate at 23.0°, 29.4°, and 39.3°, and of barium sulfate at 22.7°, 24.8°, 26.8°, and 31.5°, respectively. These observations suggest that the zinc oxide was attached to the surface of calcium carbonate and barium sulfate.

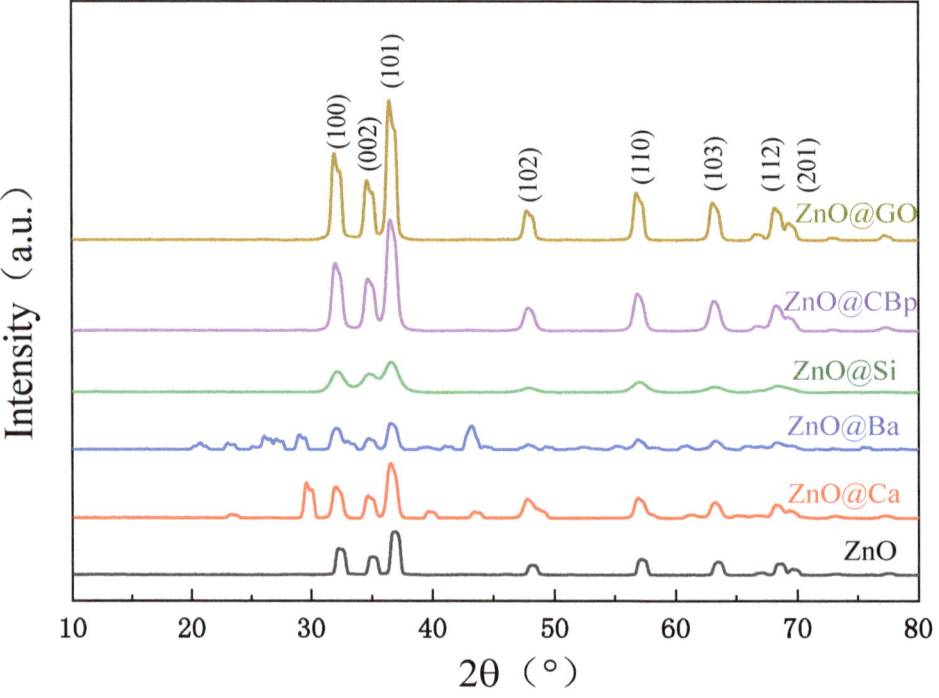

Figure 2. XRD of zinc oxide with core–shell structure of different nuclear materials.

3.1.2. SEM

The morphology of ZnO with different nucleosomal materials is presented in Figure 3. The high-resolution images of ZnO@$CaCO_3$ powder showed the formation of ZnO nanoparticles, which were spherical in size, uniform, and dense. ZnO was well coated around the nucleosomal material in Figure 3a. In Figure 3b, ZnO mainly adhered to the carrier surface in an irregular cylindrical shape, with particles adhering to each other in agglomeration. In contrast, the flaky structure of zinc oxide shown in Figure 3c,d had numerous bumpy blocks, indicating poor dispersion and slight agglomeration of zinc oxide monomers distributed on the carrier surface in a sea-urchin-like crystal form. As displayed in Figure 3e, most GO nanosheets are stacked, curled, and entangled together. The synthesized nano-ZnO exhibits an obvious tendency for the nanoparticles to agglomerate. It should be noted that the surface of GO nanosheets is covered by densely packed and irregularly shaped nano-ZnO on a large scale. Some nano-ZnO particles that are grown on the brink of the interlayer and inside the interlayer of GE nanosheets did not wrap the nucleosomal material well, and most of the graphene material was exposed.

Figure 3. SEM image of the core–shell structure zinc oxide: (**a**) CaCO$_3$; (**b**) BaSO$_4$; (**c**) SiO$_2$; (**d**) CBp; (**e**) GO.

3.1.3. TEM

TEM was performed to investigate the ZnO with core–shell structure, as shown in Figure 4. Figure 4a shows that a number of the ZnO particles were electrostatically adsorbed onto the surface of calcium carbonate solid, whereas others were free and not adsorbed. In Figure 4b, the size of the ZnO particles was between 60 and 70 nm. In Figure 4c, only a small portion of the needle-like ZnO structure was attached to the SiO$_2$ surface, whereas the ZnO morphology on the silica nucleosome material was needle-like. Figure 4d illustrates that the ZnO particles were attached to the nucleosome material in a spherical structure, and it is clear that the ZnO particles were agglomerated together. Figure 4e shows the GO nanosheets are decorated by nano-ZnO 60 nm in diameter. Notably, some nano-ZnO particles are dispersed on the surface of the wrinkled GO nanosheets, and some are covered or wrapped by thin GO nanosheets, in agreement with the SEM observations.

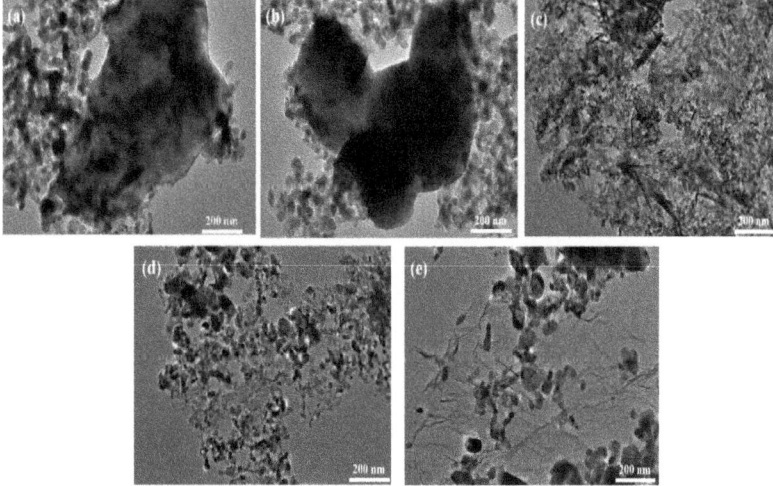

Figure 4. TEM image of the core–shell structure zinc oxide: (**a**) CaCO$_3$; (**b**) BaSO$_4$; (**c**) SiO$_2$; (**d**) CBp; (**e**) graphene.

3.2. Effect of Core–Shell ZnO of Different Core Materials on the Vulcanization Performance of Tire Tread Rubber

The vulcanization performance of ZnO with different nucleosomal materials in tire tread rubber was evaluated, as shown in Table 2. The results indicate that ZnO with a nucleosomal structure exhibits slightly lower M_H-M_L values than indirect ZnO, which may be due to its lower percentage among the same amount of ZnO, resulting in a slightly lower crosslink density. The core–shell structured zinc oxide exhibits a shorter positive vulcanization time and the fastest vulcanization rate, with a shorter scorch time compared to indirect zinc oxide. This can be attributed to the small particle size, large specific surface area, and severe lack of coordination of the core–shell structured ZnO compared to conventional ZnO, resulting in higher reactivity. In contrast, the vulcanization rate of R-ZnO@CBp and R-ZnO@GO core materials was slower due to the thermal cracking of carbon black and the easy agglomeration of graphene in the rubber, resulting in insufficient reaction between ZnO and the vulcanizing agent.

Table 2. Effect of core–shell zinc oxide with different core materials on vulcanization performance of tire tread rubber.

	R-ZnO	R-ZnO@Ca	R-ZnO@Ba	R-ZnO@Si	R-ZnO@CBp	R-ZnO@GO
M_L/dN·m	1.32	1.45	1.70	1.88	1.71	1.65
M_H/dN·m	18.56	17.80	18.68	18.99	18.62	18.55
T_{10}/s	444	372	329	319	437	446
T_{90}/s	1997	1526	1405	1343	1809	1878
M_H-M_L/dN·m	17.24	16.35	16.98	17.11	16.91	16.90

3.3. Effect of Core–Shell ZnO of Different Core Materials on the Mechanical Properties of Tire Tread Rubber

The mechanical properties of the rubber compounds are shown in Table 3 and Figure 5. It shows that the tensile strength and elongation at break of the core–shell structured ZnO specimens are greater than those of R-ZnO, probably because of the large specific surface area, better dispersion, and greater crosslinking of the core–shell structured ZnO, which exhibits excellent mechanical properties. For core–shell structured ZnO, R-ZnO@Si has better performance, which may be due to the small size and good dispersion of ZnO attached to silica, which can be effectively combined with the promoter. On the other hand, silica acts as a reinforcing system indirectly to increase the performance of the adhesive. Comparing the tearing properties, the core–shell structured ZnO exhibits generally higher tear strength than the indirect method ZnO. The high surface activity of the small particle size zinc oxide in the core–shell structured zinc oxide promotes a dense mesh structure that increases the degree of crosslinking of the rubber and limits the movement of the molecular chains. As a result, the elastic modulus in the directional direction is smaller than that in the vertical direction, hindering the crack expansion.

Table 3. Mechanical properties of zinc oxide with core–shell structure.

	R-ZnO	R-ZnO@Ca	R-ZnO@Ba	R-ZnO@Si	R-ZnO@CBp	R-ZnO@GO
Tensile strength/MPa	18.4	19.9	19.7	20.6	19.8	20.8
Elongation at break/%	389	445	432	465	436	436
Modulus at 100% strain/MPa	3.5	3.5	3.4	3.45	3.6	3.8
Modulus at 300% strain/MPa	12.6	13.2	12.8	13.0	13.7	14.0
Tear strength/N ·mm^{-1}	49.2	49.8	52.0	52.6	51.6	50.6
Hardness	75	75	75	76	77	76

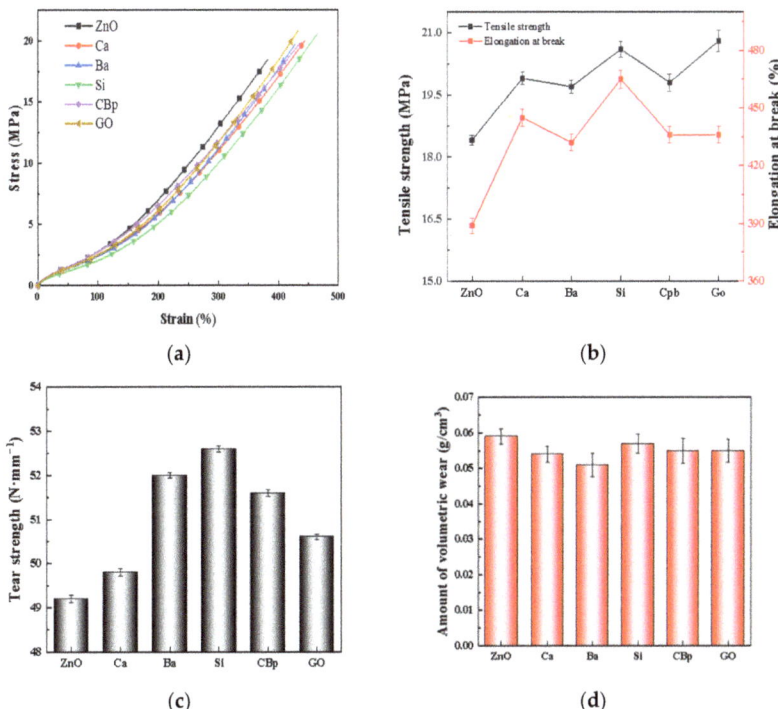

Figure 5. Mechanical properties of zinc oxide with core–shell structure. (**a**) Stress-strain curve; (**b**) tensile strength, elongation at break; (**c**) tear strength; (**d**) wear resistance.

As shown in Figure 5d, R-ZnO has the highest wear resistance, which is due to the lower crosslink density and higher wear rate, which is due to the lower degree of crosslinking and susceptibility to damage by mechanical stress. Due to the higher crosslink density of R-ZnO, the number of crosslinking sites per unit volume is higher, and the number of effective molecular chains carrying mechanical stress is higher compared to that of ZnO with a core–shell structure, resulting in a higher wear resistance. Similarly, for ZnO with a core–shell structure, the high degree of crosslinking leads to excellent wear resistance.

3.4. Effect of Core–Shell ZnO of Different Core Materials on the Dynamic Mechanical Properties of Tire Tread Rubber

The loss factors of the materials are shown in Figure 6a, and there is a significant decrease in the loss factor peak of the core–shell structured ZnO compared to the indirect method ZnO, which indicates that the core–shell structure reduces the motion of the molecular chains of the composites. tanδ at 0 °C and 60 °C then characterizes the wet-slip resistance and rolling resistance of the rubber material for tires. The loss factor curve at 0 °C shows that the wet slip resistance of the composites with the addition of ZnO with core–shell structure decreases. It indicates that the indirect method zinc oxide has stronger interaction with the matrix and higher energy loss from the movement of molecular chains. In contrast, at 60 °C, the adhesive with thermally cracked carbon black and graphene as core–shell materials has a better rolling resistance performance, probably because the core–shell structured ZnO promotes the dispersion of activator after loading, enhances the vulcanization, and builds a stronger spatial crosslinking network, which makes the hysteresis loss inside the material lower, resulting in less internal friction and lower rolling resistance. Compared with the indirect ZnO adhesive, the energy storage modulus of the adhesive with barium sulfate nucleosome material increased by 42.5%, whereas the energy storage modulus of the adhesive with graphene as the nucleosome material increased by 32.5%.

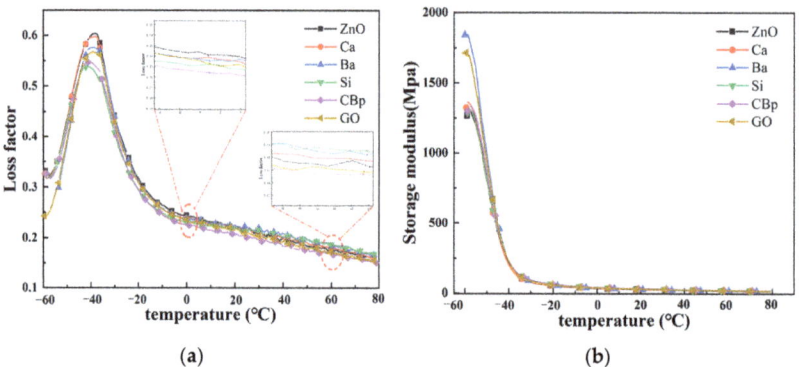

Figure 6. DMA of core–shell ZnO with different core materials. (**a**) tanδ; (**b**) storage modulus.

3.5. Effects of Core and Shell ZnO with Different Core Materials on the Aging Properties of Tire Tread Compounds

The effects of different core–shell materials of ZnO on the aging performance of rubber are shown in Figure 7. The performance of indirect method ZnO decreases more after aging; the possible reason is that the reaction between indirect method ZnO and accelerator generates less active zinc salts, which leads to low utilization of the vulcanizing agent, and the crosslinked network is dominated by polysulfide bonds, which are less thermally stable, so the performance decreases more obviously after thermal-oxidative aging. In comparison with the core–shell structure zinc oxide of different nucleosomal materials, R-ZnO@Ca and R-ZnO@Ba have better aging resistance, and the rest of the core–shell structure zinc oxide has poorer aging resistance, which may be attributed to the fact that calcium carbonate and barium sulfate core–shell structure zinc oxide in rubber are not easily agglomerated to trigger the unreacted vulcanizing agent, resulting in the crosslinking of unreacted double bonds on the rubber molecular chain; thus, the performance degradation is to a lesser extent. The difficulty of dispersion of thermally cracked carbon black and graphene in rubber makes the content of the unreacted vulcanizing agent decrease, which makes it difficult to trigger the crosslinking reaction of the double bonds on the rubber molecular chain again; i.e., the ability to resist external damage is reduced, so the degradation of rubber performance is more obvious.

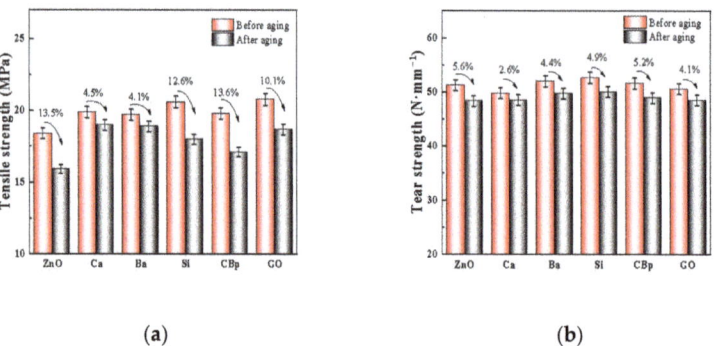

Figure 7. Effect of core–shell zinc oxide with different core materials on the aging performance of tire tread rubber. (**a**) Tensile strength after aging; (**b**) tear strength after aging.

4. Conclusions

This paper proves the core–shell structured zinc oxide was synthesized by a simple wet precipitation method and post-processing approaches without using harmful chemicals. The XRD results showed that the intense and sharp peaks in the ZnO hexagonal were highly crystalline. The SEM and TEM analysis revealed that ZnO nanoparticles (nano-ZnO) are successfully anchored onto carbonate calcium carbonate ($CaCO_3$), barium sulfate ($BaSO_4$), silicon dioxide (SiO_2), pyrolysis of carbon black (CBp), and graphene oxide (GO) sheets. The performance of NR/SBR/BR compounds was studied by adding ZnO with different nucleosomes to the formulation of semi-steel radial tire tread rubber. The results showed that the core–shell structured ZnO with low ZnO content possesses a higher vulcanization efficiency and much stronger reinforcement effect on the mechanical performance properties of NR/SBR/BR compounds compared with the indirect method zinc oxide, results of which are positively correlated with the good dispersion of ZnO throughout the NR matrix, the enhanced interfacial interaction between the ZnO and the matrix, and the high vulcanization efficiency of nano-ZnO. For the different core–shell structured zinc oxide materials, the ZnO@silica-based rubber is superior in mechanical and abrasion resistance, and the calcium-carbonate-based core–shell structured zinc oxide has excellent aging resistance. Overall, the performance of core–shell structured zinc oxide product is basically the same as that of indirect zinc oxide products, while the amount of zinc oxide in rubber can be reduced, which can also lead to a reduction in the production cost of rubber products. Accordingly, the core–shell structured ZnO with lower content is very competitive for preparing rubber composites with high performance, and it may be regarded as a substitute of conventional ZnO for application in rubber composites.

Author Contributions: Conceptualization, Z.H. and Q.C.; Data curation, Z.W. and X.L.; Formal analysis, Z.W. and H.L.; Funding acquisition, Q.C.; Investigation, Z.G.; Methodology, Z.G.; Project administration, Q.C.; Resources, H.L. and Q.C.; Software, Z.W. and X.L.; Supervision, Q.C.; Validation, H.L.; Visualization, Z.H. and Z.G.; Writing—original draft, Z.W.; Writing—review and editing, Z.H. and Q.C. All authors have read and agreed to the published version of the manuscript.

Funding: This research was funded by the National Key R&D Program of China under grant number 2022YFD2301202.

Institutional Review Board Statement: Not applicable.

Informed Consent Statement: Not applicable.

Data Availability Statement: The data that support the findings of this study are available from the corresponding author upon reasonable request.

Conflicts of Interest: The authors declare no conflict of interest.

References

1. Agarwal, H.; Kumar, S.V.; Rajeshkumar, S. A review on green synthesis of zinc oxide nanoparticles—An eco-friendly approach. *Resour. Effic. Technol.* **2017**, *3*, 406–413. [CrossRef]
2. Anand, A.; Nussana, L.; Shamaan, M.P.; Ekwipoo, K.; Sangashetty, S.G.; Jobish, J. Synthesis and Characterization of ZnO Nanoparticles and Their Natural Rubber Composites. *J. Macromol. Sci. Part B* **2020**, *11*, 5619–5621. [CrossRef]
3. Rhodes, E.P.; Ren, Z.; Mays, D.C. Zinc leaching from tire crumb rubber. *Environ. Sci. Technol.* **2012**, *46*, 12856–12863. [CrossRef] [PubMed]
4. Kołodziejczak-Radzimska, A.; Jesionowski, T. Zinc Oxide—From Synthesis to Application: A Review. *Materials* **2014**, *7*, 2833–2881. [CrossRef] [PubMed]
5. Boopasiri, S.; Thaptong, P.; Sae-Oui, P.; Siriwong, C.; Zhuravlev, M.; Sazonov, R.; Kholodnaya, G.; Pyatkov, I.; Ponomarev, D.; Konusov, F.; et al. Synthesis and characterization of zinc oxide nanopowder. *Inorg. Nano-Met. Chem.* **2021**, *51*, 798–804.
6. Gaca, M.; Pietrasik, J.; Zaborski, M.; Okrasa, L.; Boiteux, G.; Gain, O.; Magdalena, G. Effect of zinc oxide modified silica particles on the molecular dynamics of carboxylated acrylonitrile-butadiene rubber composites. *Polymers* **2017**, *9*, 645. [CrossRef] [PubMed]
7. Boopasiri, S.; Thaptong, P.; SaeOui, P.; Siriwong, C. Fabrication of zinc oxide-coated microcrystalline cellulose and its application in truck tire tread compounds. *J. Appl. Polym. Sci.* **2022**, *139*, 52701–52703. [CrossRef]
8. Thaptong, P.; Boonbumrung, A.; Jittham, P.; Sae-oui, P. Potential use of a novel composite zinc oxide as eco-friendly activator in Tire tread compound. *J. Polym. Res.* **2019**, *26*, 226. [CrossRef]

9. Jahanbakhsh, F.; Ebrahimi, B. Modified Activated Carbon with Zinc Oxide Nanoparticles Produced from Used Tire for Removal of Acid Green 25 from Aqueous Solutions. *Am. J. Appl. Chem.* **2016**, *4*, 8–13. [CrossRef]
10. Maciejewska, M.; Sowińska, A.; Kucharska, J. Organic Zinc Salts as Pro-Ecological Activators for Sulfur Vulcanization of Styrene–Butadiene Rubber. *Polymers* **2019**, *11*, 1723. [CrossRef]
11. Lin, Y.; Zeng, Z.; Zhu, J.; Chen, S.; Yuan, X.; Liu, L. Graphene nanosheets decorated with ZnO nanoparticles: Facile synthesis and promising application for enhancing the mechanical and gas barrier properties of rubber nanocomposites. *RSC Adv.* **2015**, *5*, 57771–57780. [CrossRef]
12. Kim, I.-J.; Kim, W.-S.; Lee, D.-H.; Kim, W.; Bae, J.-W. Effect of nano zinc oxide on the cure characteristics and mechanical properties of the silica-filled natural rubber/butadiene rubber compounds. *J. Appl. Polym. Sci.* **2010**, *117*, 1535–1543. [CrossRef]
13. Thomas, S.P.; Mathew, E.J.; Marykutty, C.V. Synthesis and effect of surface modified nano ZnO in natural rubber vulcanization. *J. Appl. Polym. Sci.* **2012**, *124*, 3099–3107. [CrossRef]
14. Roy, K.; Alam, M.N.; Mandal, S.K.; Debnath, S.C. Sol–gel derived nano zinc oxide for the reduction of zinc oxide level in natural rubber compounds. *J. Sol-Gel Sci. Technol.* **2014**, *70*, 378–384. [CrossRef]
15. Lee, Y.H.; Cho, M.; Nam, J.D.; Lee, Y. Effect of ZnO particle sizes on thermal aging behavior of natural rubber vulcanizates. *Polym. Degrad. Stab.* **2018**, *148*, 50–55. [CrossRef]
16. He, Q.; Zhou, Y.; Wang, G.; Zheng, B.; Qi, M.; Li, X.; Kong, L. Effects of two nano-ZnO processing technologies on the properties of rubber. *Appl. Nanosci.* **2018**, *8*, 2009–2020. [CrossRef]
17. Qin, X.; Xu, H.; Zhang, G.; Wang, J.; Wang, Z.; Zhao, Y.; Wang, Z.; Tan, T.; Bockstaller, M.R.; Zhang, L.; et al. Enhancing the Performance of Rubber with Nano ZnO as Activators. *ACS Appl. Mater. Interfaces* **2020**, *12*, 48007–48015. [CrossRef]
18. Hasany, S.F.; Hussain, S.; Usman Ali, S.M.; Abdul-Kadhim, W.; Amir, M. ZnO Nanostructures: Comparative synthetic and characterization studies. *Micro Nano Lett.* **2020**, *15*, 972–976. [CrossRef]
19. Awad, M.A.; Ibrahim, E.; Ahmed, A.M. Synthesis and thermal stability of ZnO nanowires. *J. Therm. Anal. Calorim.* **2014**, *117*, 635–642. [CrossRef]
20. Alam, M.N.; Kumar, V.; Park, S.-S. Advances in Rubber Compounds Using ZnO and MgO as Co-Cure Activators. *Polymers* **2022**, *14*, 5289. [CrossRef]
21. Ikeda, Y.; Yasuda, Y.; Ohashi, T.; Yokohama, H.; Minoda, S.; Kobayashi, H.; Honma, T. Dinuclear Bridging Bidentate Zinc/Stearate Complex in Sulfur Cross-Linking of Rubber. *Macromolecules* **2015**, *48*, 462–475. [CrossRef]
22. Yang, Z.; Huang, Y.; Xiong, Y. A functional modified graphene oxide/nanodiamond/nano zinc oxide composite for excellent vulcanization properties of natural rubber. *RSC Adv.* **2020**, *10*, 41857–41870. [CrossRef] [PubMed]
23. Suntako, R. Influence of Zinc Oxide Nanograins on Properties of Epoxidized Natural Rubber Vulcanizates. *Key Eng. Mater.* **2017**, *4534*, 79–87. [CrossRef]
24. Suntako, R. Effect of synthesized ZnO nanoparticles on thermal conductivity and mechanical properties of natural rubber. *IOP Conf. Ser. Mater. Sci. Eng.* **2018**, *284*, 25–31. [CrossRef]
25. Mohammed, S.; Alhumdany, A.; Al Waily, M. Effect of nano zinc oxide on tensile properties of natural rubber composite. *Kufa J. Eng.* **2018**, *9*, 77–90. [CrossRef]
26. Hayeemasae, N.; Rathnayake, W.G.I.U.; Ismail, H. Effect of ZnO Nanoparticles on the Simultaneous Improvement in Curing and Mechanical Properties of NR/ Recycled EPDM Blends. *Prog. Rubber Plast. Recycl. Technol.* **2018**, *34*, 1–18. [CrossRef]
27. Hadi, F.A.; Kadhim, R.G. A Study of the Effect of Nano Zinc Oxide on Cure Characteristics and Mechanical Properties of Rubber Composites. *J. Phys. Conf. Ser.* **2019**, *1234*, 13–18. [CrossRef]
28. Roy, K.; Alam, M.N.; Mandal, S.K.; Debnath, S.C. Surface modification of sol–gel derived nano zinc oxide (ZnO) and the study of its effect on the properties of styrene–butadiene rubber (SBR) nanocomposites. *J. Nanostruct. Chem.* **2014**, *4*, 133–142. [CrossRef]
29. Li, Y.; Sun, H.; Zhang, Y.; Xu, M.; Shi, S.Q. The three-dimensional heterostructure synthesis of ZnO/cellulosic fibers and its application for rubber composites. *Compos. Sci. Technol.* **2019**, *177*, 10–17. [CrossRef]
30. Gherekhlo, Z.D.; Motiee, F.; Khorrami, S.A. Synthesis of the CoO.CaO/ZnO core-shell nanopigment and investigation of its effects on the properties of rubber compounds based on the acrylonitrile butadiene elastomer (NBR). *IIOAB J.* **2016**, *7*, 394–400.
31. Sabbagh, F.; Kiarostami, K.; Mahmoudi Khatir, N.; Rezania, S.; Muhamad, I.I. Green Synthesis of Mg0.99 Zn0.01O Nanoparticles for the Fabrication of κ-Carrageenan/NaCMC Hydrogel in order to Deliver Catechin. *Polymers* **2020**, *12*, 861. [CrossRef] [PubMed]

Disclaimer/Publisher's Note: The statements, opinions and data contained in all publications are solely those of the individual author(s) and contributor(s) and not of MDPI and/or the editor(s). MDPI and/or the editor(s) disclaim responsibility for any injury to people or property resulting from any ideas, methods, instructions or products referred to in the content.

Article

Enhancing Natural Rubber Tearing Strength by Mixing Ultra-High Molecular Weight Polyethylene Short Fibers

Jun He [1], Baoyuan Huang [2], Liang Wang [1], Zunling Cai [1], Jing Zhang [1,*] and Jie Feng [1,*]

[1] College of Materials Science and Engineering, Zhejiang University of Technology, Hangzhou 310014, China
[2] Linhai Weixing New Construction Materials Co., Ltd., Taizhou 317016, China
* Correspondence: zhangjing@zjut.edu.cn (J.Z.); fengjie@zjut.edu.cn (J.F.)

Abstract: Rubber products generally need to have high resistance to abrasion, tear, and cutting. Filling short fiber with strong mechanical properties and forming a net in the rubber matrix is a good method to realize the above aims. In this article, ultra-high molecular weight polyethylene (UHMWPE) short fibers with a diameter of 20 μm and a length of 2 cm were filled into natural rubber (NR) to improve the tear strength of the NR. The influence of the short fiber mass fraction and vulcanization conditions on the mechanical properties of the composites were investigated. The results show that the milling process and vulcanization conditions are key factors in enhancing tear resistance performance. Double-roll milling and vulcanization at 143 °C for 40 min result in strong interfacial adhesion between the UHMWPE short fibers and the NR. The addition of 2 phr of UHMWPE fiber increases the tear strength of the composite material by up to 150.2% (from 17.1 kN/m to 42.8 kN/m) while also providing excellent comprehensive performance. Scanning electron microscope (SEM) imaging confirmed that the UHMWPE short fibers are dispersed in the NR matrix homogeneously, and the interface is close and compact. As a control experiment, UHMWPE resin powder was directly filled into the NR, and then the composite was vulcanized using the same process as that used for the NR/UHMWPE short fiber composite. The results show that the mechanical strength of the NR/resin powder composite exhibits minor improvement compared with NR. As there is no complicated surface modification of the UHMWPE fiber, the results reported may be helpful in improving the tear resistance of the industrially prepared rubber conveyor belts.

Keywords: UHMWPE; short fiber; natural rubber; conveyor belt; tear resistance

Citation: He, J.; Huang, B.; Wang, L.; Cai, Z.; Zhang, J.; Feng, J. Enhancing Natural Rubber Tearing Strength by Mixing Ultra-High Molecular Weight Polyethylene Short Fibers. *Polymers* 2023, 15, 1768. https://doi.org/10.3390/polym15071768

Academic Editors: Yuwei Chen and Yumin Xia

Received: 26 January 2023
Revised: 24 March 2023
Accepted: 29 March 2023
Published: 1 April 2023

Copyright: © 2023 by the authors. Licensee MDPI, Basel, Switzerland. This article is an open access article distributed under the terms and conditions of the Creative Commons Attribution (CC BY) license (https://creativecommons.org/licenses/by/4.0/).

1. Introduction

Rubber conveyor belts are some of the most important industrial transportation tools in use today, and so are required to have high strength, flame retardancy, and high resistance to abrasion, high temperature, fatigue, tear, and impact [1]. Among these properties, due to severe working conditions, the conveyor belt especially needs to have tear resistance [2] and impact resistance [3]. Recently, differing from the usual inorganic fillers that are employed, the use of fibers for reinforcing rubber has shown great potential. In the last decades, short fiber/rubber composites (SFRC) have been studied in combination with various types of fibers due to their superior physical and chemical properties [4], such as glass fiber [5], carbon fiber [6], ultra-high molecular weight polyethylene (UHMWPE) fiber [7], aramid fiber [8], and natural fiber [9–11].

Among all the types of short fibers, carbon fiber, UHMWPE fiber, and aramid fiber may be the three most well-used fibers for improving the mechanical performance of rubber. The modulus of carbon fiber is far higher than that of rubber, so its reinforcing effect is not particularly high [12], although there is good compatibility between the two substances. Aramid fiber is a polar fiber, and so has poor compatibility with most non-polar rubbers [13]. On the contrary, the modulus of UHMWPE fiber is well-matched with non-polar rubbers and exhibits good intrinsic compatibility. Moreover, it also has a low density, high tensile strength, and high resistance to abrasion and thus is expected to exhibit good performance in rubbers [14].

The effect of filling untreated or chromic-acid-treated UHMWPE short fibers (the length is 0.5 cm) on natural rubber (NR) was investigated by Li et al. [15]. The tear strength of the fiber/NMR composite consisting of untreated or treated fiber increased no more than 30% compared with that of NR alone. Zhang et al. [16] treated UHMWPE fibers with ozone and then UV grafting glycidyl methacrylate (GMA) in order to enhance the interfacial properties of the fiber/rubber composites. The result showed that by adding an amino-containing adhesive RA reagent to the formula, the adhesive force between the fibers and SBR increased by 79% over that of the untreated fibers. However, in their study, no tensile and tear strengths for the fiber/rubber composites were reported. Tu et al. [17] constructed a polydopamine (PDA) functionalization platform and then deposited zinc oxide nanoparticles on the UHMWPE fiber surface and found that the interfacial adhesion with the rubber matrix was enhanced by 85.4%. Later, they [18] proposed a lower-cost surface modification strategy by replacing expensive dopamine with the catechol/tetraethylenepentamine two-component system.

Although further surface modification to the UHMWPE fiber has provided better interfacial performance between the fibers and the non-polar rubber, these modifications are complicated and may not be cost-effective [19,20]. In fact, the UHMWPE fibers have good compatibility with most non-polar rubbers because they have similar chemical groups. Further surface modification to the UHMWPE fiber may be unnecessary except for the simple formation of a covalent bond at the interface. On the contrary, the length of the UHMWPE short fiber and the vulcanization process may be more beneficial in increasing the tear strength and other mechanical properties of the rubber.

In the practical production of rubber products such as jugged triangular belts, aramid short fibers can be filled into the rubber in order to avoid stress cracking at the bottom of such structures. The length of such aramid short fibers lies in the region of 1~3 mm and can often be shorter than 1 mm. However, the cost of the aramid short fibers is particularly expensive (250–400 RMB per kg), and so few industrial-scale plants are willing to use them. Compared with aramid short fiber, UHMWPE short fiber is cost-effective (150–250 RMB/kg). If UHMWPE could be homogeneously dispersed in the rubber matrix, a reinforcing effect may be obtained, similar to the function provided by the aramid short fibers. Additionally, fibers that are longer in length will help improve the tear strength of the resulting composite more obviously.

Based on promoting the practical application of the UHMWPE short fibers in the rubber industry, in this work, complex and expensive physical or chemical modifications to the fiber surface have neither been used. On the contrary, different amounts of original UHMWPE short fibers with lengths of 2 cm were homogeneously dispersed in NR by double-roll milling at a certain temperature. The mixture was then vulcanized at a specific temperature. The effects of the fiber amount, the mixing, and the vulcanization process on the mechanical performance of the rubber were studied systemically. The results showed that under vulcanization conditions of 143 °C for 40 min, The tearing performance of the NR is increased by up to 150.2% following the addition of 2 phr of UHMWPE fiber. This preparation method does not require any surface modification of the UHMWPE fiber; thus, the results of this study may be beneficial to the rubber conveyor belt industry.

2. Experimental Section

2.1. Materials

The NR and other chemical reagents (industrial grade) were obtained from Zhejiang Fenfei Rubber & Plastic Products Co., Ltd. (Taizhou, Zhejiang, China). The UHMWPE short fibers were purchased from Zhejiang Qianxilong Special Fiber Co., Ltd. (Jinhua, Zhejiang, China). The diameter and the length of the short fibers were approximately 20 μm and 2 cm, respectively. The master batch of the NR was prepared by mixing NR, carbon black

(CB), aromatic oil (oil), zinc oxide (ZnO), stearic acid (SA), sulfur, accelerator (TBBS), and UHMWPE short fibers. The specific formula of the composite is listed in Table 1.

Table 1. Compounding formulation of NR/UHMWPE short fibers.

Sample	NR	CB	Oil	ZnO	SA	Sulfur	TBBS	UHMWPE Short Fiber/Phr
1	100	0	6	5	3	2.2	0.8	0
2	100	0	6	5	3	2.2	0.8	2
3	100	0	6	5	3	2.2	0.8	4
1 *	100	50	6	5	3	2.2	0.8	0
2 *	100	50	6	5	3	2.2	0.8	2
3 *	100	50	6	5	3	2.2	0.8	4

2.2. Preparation of NR/UHMWPE Short Fiber Composites

The components listed in Table 1 were dried in an electric thermostatic drying oven at 60 °C for 12 h and weighed according to the formulation. The NR and short fibers were mixed on a two-roll mill (LRM-S-150/3E, Labtech Engineering Co., Ltd., Bangkok, Thailand) over a three-step process. First, rubber was mixed with CB, aromatic oil, ZnO, and SA at 80 °C with a speed of 15/12 rpm (front/rear roll). After 3 min, the UHMWPE short fibers were added and milled for a further 10 min. The shear force generated by the rolls leads to most short fibers being oriented in the rolling direction, as reported by Andideh M [21]. Finally, the NR/fiber composites were obtained after sulfur and accelerant (TBBS) were incorporated into the system by mixing for another 3 min. In the two later steps, the milling condition is the same as that of the first step.

Due to the macroscopic length of the UHMWPE short fiber (2 cm), the oriented fibers presenting on the composite sample surface can be seen clearly. In addition to using a two-roll milling process, a closed mixer was also applied in order to mix the components listed in Table 1 for future industrial manufacture. The temperature was 105 °C, the rotor speed was set at 22 rpm, and the mixing time was 10 min. The fibers were mixed homogeneously into the NR using this process. However, the composites that were mixed with the closed mixer still needed to be formed into sample pieces by two-roll milling. During the two-roll milling process, the fibers were again orientated along the direction of the roll rotation. Thus, for most samples in this study, the fibers were only mixed into NR via the two-roll milling process.

The creep and melting temperature of the UHMWPE short fibers are close to the vulcanization temperature of general rubber. Therefore, in order to determine the optimized vulcanization temperature, the fiber/NR composite sheets with 2 mm thickness were firstly vulcanized under a thermal pressing machine (LP-S-50, Labtech Engineering Co., Ltd., Bangkok, Thailand) at 140 °C for 40 min, 143 °C for 40 min and 150 °C for 30 min, respectively. Additionally, all the specimens were stored for 24 h at room temperature before further testing. Next, the samples consisting of different fiber content were studied based on the optimal vulcanization process. To further demonstrate the function of the UHMWPE short fiber, UHMWPE resin powder (Shanghai Lianle Chemical Co., Ltd., Shanghai, China, mesh number 100) was directly filled into the NR, and the composites were vulcanized using the same process as used for the NR/short fiber composite.

2.3. Measurements and Characterization

2.3.1. Thermal Property of the UHMWPE Short Fiber

Irrespective of using milling or vulcanization with NR, the thermal performance of the UHMWPE short fiber should be investigated; otherwise, their excellent mechanical properties would disappear as soon as the temperature exceeds the fiber's creep or melting temperatures. Thus, differential scanning calorimetry (DSC, Mettler Toledo, Zurich, Switzerland) was used to characterize the thermal behavior of the fiber under a nitrogen atmosphere and heated at 10 °C/min from room temperature to 200 °C.

2.3.2. Curing Characteristics of the Fiber/NR Composite

In order to ensure a suitable vulcanization time for the composite, the NR samples without the UHMWPE short fiber filling were vulcanized at 143 °C and 150 °C by a rheometer (M-3000A, Gotech testing machines Inc., Taiwan, China) until the T90 was measured. The curing characteristic data of the fiber/NR composites were determined using the same rheometer at a temperature of 143 °C. The pressure for vulcanization was 7.5 MPa.

2.3.3. Mechanical Properties of the Fiber/NR Composite

The tensile properties and tear strength of the cured fiber/NR composite samples were determined by an electronic universal testing machine (5966, instron, Boston, MA, USA) based on ISO 37-2005 (Type 2) and ISO 34-1-2010 (trouser test piece), respectively. For the conveyor belt, horizontal tearing is more problematic, especially longitudinal tearing (along the running direction). The test was performed perpendicular to the orientation of the fiber. A Shore hardness tester (LX-A, Shanghai Precision Instruments Co., Ltd., Shanghai, China) was used to measure the hardness under ASTM-D2240 conditions. In each test, 5 replicas of specimens were used, and the average and standard deviation of each test was given in the form of a table.

The difference inelasticity difference between the fiber and the NR resulted in disconnection when the composite was stretched too long; the application of UHMWPE resin powder was expected to result in the formation of shorter fibers in situ during the filling process. NR, UHMWPE resin powder, and additives were mixed using the same processing as used for the preparation of the NR/UHMWPE short fiber composite. Further, 2 phr, 4 phr, and 6 phr resin powder were filled into the NR, respectively. Then the composites were vulcanized at 143 °C for 40 min.

2.3.4. Micromorphology Analysis

In order to observe the dispersement of the short fibers in the NR matrix, the fiber/NR composite specimens were quenched in liquid nitrogen, and a section of the surface was coated with platinum for 45 s in a sputter coater (Sputter Coater 108, Cressington Scientific Instruments Ltd., Watford, UK). Finally, the fracture surface was observed with a scanning electron microscope (SEM, VEGA 3, Tescan, Czech Republic) at 15 kV. The interface between the fiber and the NR was analyzed to determine the compatibility between the UHMWPE short fiber and the NR.

3. Results and Discussion

3.1. DSC of UHMWPE Short Fiber

The thermal characteristics of the UHMWPE short fiber are shown in Figure 1. It can be seen that the fiber starts melting at approximately 138.6 °C. Furthermore, the melting peak is approximately 146.3 °C; this means that the fiber will lose its excellent mechanical performance once the temperature exceeds 146 °C due to macromolecular disorientation [22]. Thus, the vulcanization temperature of the NR/Fiber composite must be below 146.3 °C and even below 138.6 °C in order to guarantee excellent mechanical performance. However, considering the efficiency of the vulcanization process, three temperatures, e.g., 140 °C, 143 °C, and 150 °C, were still employed as the possible curing temperatures.

The UHMWPE short fiber has poor heat deformation resistance compared with the aramid short fiber. Cross-linking between the orientated macromolecules of the fiber can improve the resistance to heat deformation. However, simultaneous control of orientation and cross-linking is difficult; cross-linking at a later time, e.g., cross-linking after orientation completion, could provide good resistance to heat deformation or disorientation. Cross-linking by radiation may be competent if the irradiation dosage is not too large. Another later cross-linking may be realized by mixing HDPE being grafted with siloxane before fiber formation and then slowly curing by moisture in the air. A conservative vulcanization temperature (140 °C or 143 °C) for the UHMWPE short fiber with a melting peak at approximately 146.3 °C, compared to 150–160 °C, may provide better results.

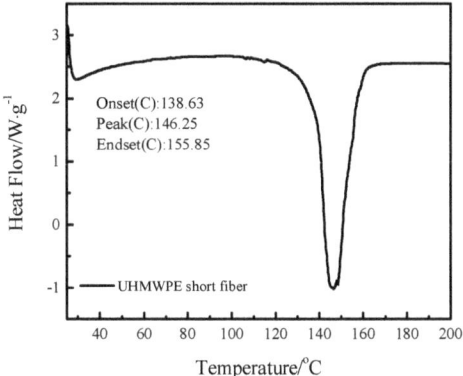

Figure 1. The DSC curves of the UHMWPE short fiber.

3.2. Curing Characteristics of the NR and the UHMWPE Fiber/NR

The curing performances of the rubber compounds are shown in Table 2. These results show that the addition of the fiber had an insignificant effect on the dynamics of the vulcanization. It appears that the vulcanization period can be reduced significantly by filling with CB. However, when TS1 is deduced, it can be seen that the vulcanization periods are not obviously different, and they are all in the 16–23 min range. However, in order to ensure sufficient vulcanization, the optimized vulcanization process temperature was set at 143 °C over 40 min. In order to obtain higher vulcanization efficiency, a temperature of 150 °C over 30 min was also investigated.

Table 2. Vulcanization data of NR/Fiber compounds at 143 °C.

Sample	M_L/N·m	M_H/N·m	Ts1/min	T_{90}/min
1	0.97	6.41	25.68	44.26
2	0.64	6.84	24.10	42.27
3	0.52	5.98	25.67	45.99
1 *	2.75	20.38	10.61	27.95
2 *	2.50	22.05	9.66	26.20
3 *	2.26	18.45	10.39	33.64

3.3. Mechanical Properties of the UHMWPE Fiber/NR Composites

Due to the UHMWPE fiber being impressionable to temperature, three different vulcanization temperatures were studied. The fiber/NR composites with different fiber contents were first vulcanized at 140 °C for 40 min to avoid disorientation of the UHMWPE fiber at high temperatures (Table 3). The 2 phr fiber filling was shown to be the best performing, especially with regard to elongation at break. Table 4 lists the mechanical properties of the composites filled with 2 phr fiber vulcanized under different conditions. For samples with the same 2 phr fiber filling, the best vulcanization process had a temperature of 143 °C and a time of 40 min with regards to the tensile stress at 300%, break, and tear strength. Complete vulcanization at temperatures lower than 143 °C may take longer; a temperature of 150 °C results in the disorientation of the fiber.

Table 3. Mechanical properties of fiber/NR composites with different fiber contents.

Sample	Vulcanization	Tensile Stress at 300% (MPa)	Elongation at Break (%)	Tensile Stress at Break (MPa)	Tear Strength (kN/m)
1 *	140 °C, 40 min	4.97 ± 0.30	1035 ± 31	25.38 ± 0.30	13.05 ± 0.74
2 *	140 °C, 40 min	9.70 ± 0.76	628 ± 23	18.77 ± 0.46	32.41 ± 1.21
3 *	140 °C, 40 min	12.27 ± 1.8	412 ± 71	15.19 ± 0.45	50.36 ± 2.99

Note: * means sample filling with CB, please see Table 1 (the same as following tables or figures).

Table 4. Mechanical properties of fiber/NR composites vulcanized under different conditions.

Sample	Vulcanization	Tensile Stress at 300% (MPa)	Elongation at Break (%)	Tensile Stress at Break (MPa)	Tear Strength (kN/m)
2 *	140 °C, 40 min	9.70 ± 0.76	628 ± 23	18.77 ± 0.46	32.41 ± 1.21
2 *	143 °C, 40 min (closed mixing)	10.65 ± 2.2	652 ± 24	18.85 ± 0.83	40.92 ± 2.30
2 *	150 °C, 30 min	9.26 ± 1.2	625 ± 36	20.20 ± 0.55	31.86 ± 2.68

The disorientation of the fiber in the composite intrinsically decreases the reinforcing effect of the fibers in the NR. It must be noted that the composite with 2 phr fiber was prepared in a closed mixing machine; it was still pressed into one piece by a two-roll mill. Thus, in the following sample preparation, only a two-roll mill was used. Moreover, the typical vulcanization of rubber products is 150 °C for 10~15 min. A temperature of 143 °C and a 40 min time period are not effective for practical vulcanization. Maybe a longer vulcanization period would bring better mechanical performance. However, formulas matching low-temperature vulcanization, e.g., at 140 °C for 10~15 min, will be conducted in future studies. For example, an accelerator with higher activity or a new vulcanization agent should be used if the temperature is limited blow 150–160 °C.

The stress–strain curves of the fiber/NR composites are shown in Figure 2. The mechanical properties of the composites are listed in Table 5. The pure NR shows the maximum elongation at break, minimum tensile strength at 300%, and tear strength. The NR/CB samples have greater hardness, tensile strength at 300%, and tear strength compared with the corresponding samples without a CB filling. As the amount of fiber increases, the elongation at break and tensile strength at break both decrease. However, the hardness, tensile stress at 300%, and tear strength all increases significantly. The decrease in elongation and tensile strength at break may be caused by cavitation or disconnection during stretching, which occurs at the interface between the fiber surface and the NR matrix [23]. The lack of a covalent bond and different elasticities between the fiber and the NR is most likely responsible for such cavitation or disconnection.

Table 5. Mechanical properties of fiber/NR composites with different formulas.

Sample	Hardness (HA)	Tensile Stress at 300% (MPa)	Elongation at Break (%)	Tear Strength (kN/m)
1	39 ± 0.7	0.82 ± 0.03	1680 ± 95	9.5 ± 1.2
2	44 ± 1.2	3.15 ± 0.28	1239 ± 41	14.7 ± 3.5
3	46 ± 1.0	5.25 ± 0.50	891 ± 73	18.6 ± 2.6
1 *	63 ± 0.6	6.49 ± 0.13	912 ± 53	17.1 ± 0.9
2 *	75 ± 0.9	11.42 ± 1.5	576 ± 41	42.8 ± 3.0
3 *	77 ± 0.8	14.64 ± 1.9	484 ± 66	64.9 ± 4.1

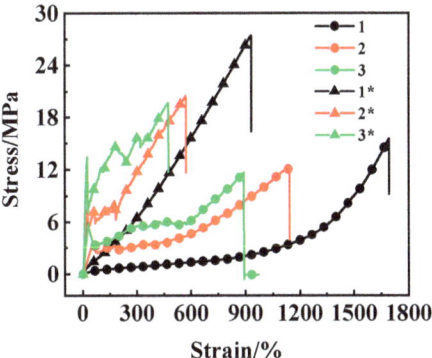

Figure 2. Stress–strain curves of NR and UHMWPE fiber/NR composites. The samples were mixed with the two-roll milling process, and the vulcanization process is 143 °C over 40 min.

The covalent bond between the fiber surface and the NR can be formed by treating the fiber using plasma or corona in order to form the active chemical group, such as hydroxyl substituents. A coupling agent is then grafted with the terminal chemical group, which can take part in the cross-linking reaction of the rubber. In our earlier studies, we treated hydrophilic silica nanoparticles with the same method and enhanced the interaction between the particles and the rubber matrix and the mechanical properties of the reinforced rubber [24,25]. The difference in elasticity between the fiber and the NR can be adjusted by shortening the length of the fiber, i.e., to 1~3 mm. In practical productions, the tear strength of the rubber can be improved using aramid short fiber with a length of 1~3 mm. This area of work will be reported in future studies.

The fibers form a net in the NR matrix when the strain is low; this prevents the expansion of the notch when being torn in a perpendicular direction (Figure 3). This phenomenon indicates that there is significant interfacial interaction between the fiber and the rubber, and so a greater strength is required to pull the fibers out of the rubber matrix. The same surface hydrophobic property of the fiber and the NR is likely responsible for the good interfacial compatibility. At a high strain, the UHMWPE fiber net is damaged due to the differing elasticity or deformability between the NR and the fiber; the delaminated fibers work as stress concentrators, which results in premature failure of the composite [26]. This suggests that the composite introduced in this work may be valuable at a strain of less than 450%.

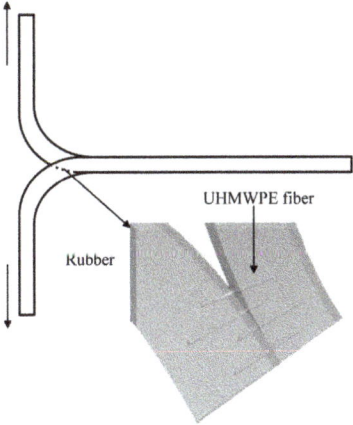

Figure 3. Schematic diagram of the mechanism for improving the tearing strength of NR.

In fact, the breaking sound caused by the disconnection of the fiber from the NR matrix could obviously be heard when the composite was stretched too much, e.g., exceeded 450%. Fortunately, many applications of rubber conveyor belts do not require high resistance to strain because the skeleton enhancement layer cannot be stretched too long. It is possible that the cover layer rubber requires high deformability. However, higher elongation at break may be possible by using shorter fibers, i.e., with lengths of 1~3 mm. However, the disconnection of the fiber from the NR matrix at high strain could be avoided by the introduction of a covalent bond between the fiber surface and the NR. In tires and rubber conveyor belts, the steel wire or the textile which are employed to reinforce the rubber are typically treated with copper or adhesion to ensure they can be firmly combined with the NR.

3.4. Dispersement of the UHMWPE Short Fibers in the NR Matrix

Although the naked eye can clearly see that the UHMWPE fibers are homogeneously dispersed on the NR matrix, SEM was still used to observe the interface between the fiber and the NR matrix. The results (Figure 4) show that the fibers are well dispersed in the NR matrix vulcanized at 143 °C for 40 min and that the addition of 4 phr of fiber may be too much. The fibers show no obvious curling and creep behavior; thus, their mechanical properties are not changed significantly. By focusing on each individual fiber, it can be seen that the interface between the fiber and the NR matrix is not sharp (as the arrows show in b,d). This proves that the rubber and UHMWPE molecular chains are entangled together at the micro level.

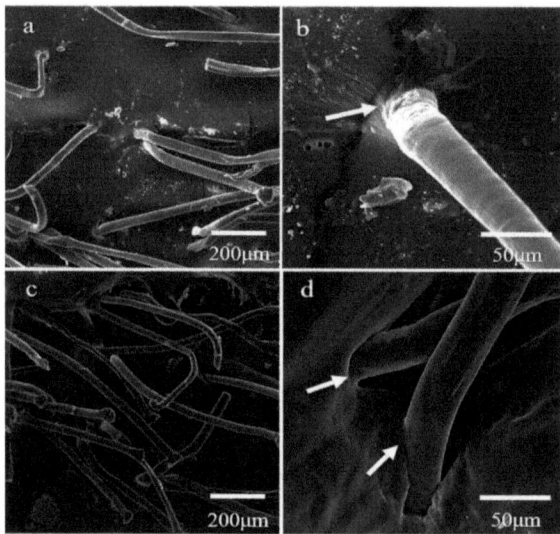

Figure 4. SEM images of the cross-section of the NR filling with 2 phr (**a,b**) and 4 phr (**c,d**) of UHMWPE short fibers (2 cm length).

Anisotropic dispersement of the short fiber in the NR matrix (when all the fibers are orientated along the running direction of the two-roll mill) is obvious when the composite is pressed into a thin piece by two-roll milling. Direct use of the closed mixer can avoid such anisotropy; however, for the formation of the thin piece, two-roll milling and orientation of the fibers are unavoidable. Such anisotropy of the short fiber may be avoided by using a thicker piece, especially when the length of the fiber is shorter. Another strategy for avoiding anisotropy is the perpendicular overlaying of multiple thin pieces.

3.5. NR/UHMWPE Resin Powder Composite

The mechanical performance is shown in Figure 5. It can be seen that the trend, which is the same as that shown in Figure 2, is observed, e.g., the higher the content of the resin powder, the lower the strength and elongation at break. Moreover, compared with the NR/fiber composite, the NR/powder composites have a higher elongation at break. This is likely because the applied powder only acts as a type of inert filler and so may not have been stretched into the fiber. This is further evidenced by the tear strengths and the respective changes observed following an increase in powder content (Table 6). Compared with the NR/short fiber composites, the NR/powder composite has a much lower tear strength.

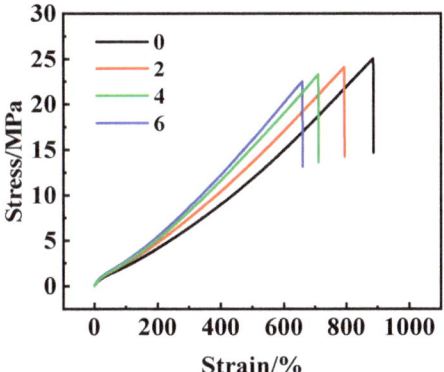

Figure 5. Stress–strain curves of NR/UHMWPE resin powder composites.

Table 6. Mechanical properties of NR/UHMWPE resin powder composites.

Sample	Hardness (HA)	Tensile Stress at 300% (MPa)	Elongation at Break (%)	Tensile Stress at Break (MPa)	Tear Strength (kN/m)
0	63 ± 0.6	6.41 ± 0.9	886 ± 13	25.06 ± 1.2	16.9 ± 1.5
2	67 ± 0.7	7.47 ± 1.3	793 ± 21	24.03 ± 0.5	17.5 ± 0.8
4	69 ± 0.3	8.22 ± 1.9	710 ± 19	23.28 ± 0.7	18.8 ± 1.0
6	70 ± 0.5	8.68 ± 1.1	659 ± 14	22.50 ± 1.3	20.5 ± 0.7

Note: Sample 0, 2, 4, 6 means the phr content of the resin powder in the NR.

4. Conclusions

In this work, UHMWPE short fibers possessing excellent mechanical properties were dispersed in the NR matrix in order to improve the tear resistance of the NR. Using the two-roll milling process and a suitable vulcanization temperature that is as high as possible but lower than the melting temperature of the UHMWPE (146.3 °C), UHMWPE short fibers/NR composites with significantly enhanced tear resistance and improved tensile stress at 300% elongation, were prepared. Among these composites, the composites consisting of 2 phr fibers exhibit the best comprehensive performance. The tear resistance was improved from 17.1 kN/m to 42.8 kN/m. SEM results indicated that the fibers are well-dispersed and show good interfacial performance with the NR matrix. Although the fiber surface has not been treated, they have natural compatibility with the rubber matrix. This work is expected to provide insight into the industrial preparation of conveyor belts. For example, the use of covalent bonds between the fiber surface and rubber, as well as the use of fibers with a shorter length, will allow for composites with better performance to be prepared.

Author Contributions: Methodology, J.H. and B.H.; Validation, Z.C.; Investigation, L.W., J.Z. and J.F.; Writing—original draft, J.H. All authors have read and agreed to the published version of the manuscript.

Funding: This research was funded by Zhejiang Provincial Key Research and Development Program (2023C01093).

Institutional Review Board Statement: Not applicable.

Informed Consent Statement: Not applicable.

Data Availability Statement: Data are contained within the article.

Conflicts of Interest: The authors declare that there is no conflict of interest.

References

1. Andrejiova, M.; Grincova, A.; Marasova, D. Failure analysis of the rubber-textile conveyor belts using classification models. *Eng. Fail. Anal.* **2019**, *101*, 407–417. [CrossRef]
2. Andrejiova, M.; Grincova, A.; Marasova, D. Measurement and simulation of impact wear damage to industrial conveyor belts. *Wear* **2016**, *368*, 400–407. [CrossRef]
3. Fedorko, G.; Molnar, V.; Grincova, A.; Dovica, M.; Toth, T.; Husakova, N.; Taraba, V.; Kelemen, M. Failure analysis of irreversible changes in the construction of rubber-textile conveyor belt damaged by sharp-edge material impact. *Eng. Fail. Anal.* **2014**, *39*, 135–148. [CrossRef]
4. Roy, K.; Debnath, S.C.; Potiyaraj, P. A critical review on the utilization of various reinforcement modifiers in filled rubber composites. *J. Elastomers Plast.* **2020**, *52*, 167–193. [CrossRef]
5. Hashemi, S.J.; Sadooghi, A.; Rahmani, K.; Nokbehrosta, S. Experimental determining the mechanical and stiffness properties of natural rubber FRT triangle elastic joint composite reinforcement by glass fibers and micro/nano particles. *Polym. Test.* **2020**, *85*, 106461. [CrossRef]
6. Tian, X.L.; Han, S.; Zhuang, Q.X.; Bian, H.G.; Li, S.M.; Zhang, C.Q.; Wang, C.S.; Han, W.W. Surface Modification of Staple Carbon Fiber by Dopamine to Reinforce Natural Latex Composite. *Polymers* **2020**, *12*, 988. [CrossRef]
7. Dong, C.L.; Shi, L.C.; Li, L.Z.; Bai, X.Q.; Yuan, C.Q.; Tian, Y. Stick-slip behaviours of water lubrication polymer materials under low speed conditions. *Tribol. Int.* **2017**, *106*, 55–61. [CrossRef]
8. Pittayavinai, P.; Thanawan, S.; Amornsakchai, T. Comparative study of natural rubber and acrylonitrile rubber reinforced with aligned short aramid fiber. *Polym. Test.* **2017**, *64*, 109–116. [CrossRef]
9. Roy, K.; Debnath, S.C.; Tzounis, L.; Pongwisuthiruchte, A.; Potiyaraj, P. Effect of Various Surface Treatments on the Performance of Jute Fibers Filled Natural Rubber (NR) Composites. *Polymers* **2020**, *12*, 369. [CrossRef]
10. Moonart, U.; Utara, S. Effect of surface treatments and filler loading on the properties of hemp fiber/natural rubber composites. *Cellulose* **2019**, *26*, 7271–7295. [CrossRef]
11. Shen, Z.; Song, W.H.; Li, X.L.; Yang, L.; Wang, C.Y.; Hao, Z.; Luo, Z. Enhancing performances of hemp fiber/natural rubber composites via polyhydric hyperbranched polyester. *J. Polym. Eng.* **2021**, *41*, 404–412. [CrossRef]
12. Sathi, S.G.; Jeon, J.; Kim, H.H.; Nah, C. Mechanical, morphological and thermal properties of short carbon and aramid fibres-filled bromo-isobutylene-isoprene rubber vulcanised with 4, 4′ bis(maleimido)diphenylmethane. *Plast. Rubber Compos.* **2019**, *48*, 115–126. [CrossRef]
13. Lin, G.Y.; Wang, H.; Yu, B.Q.; Qu, G.K.; Chen, S.W.; Kuang, T.R.; Yu, K.B.; Liang, Z.N. Combined treatments of fiber surface etching/silane-coupling for enhanced mechanical strength of aramid fiber-reinforced rubber blends. *Mater. Chem. Phys.* **2020**, *255*, 123486. [CrossRef]
14. Chhetri, S.; Bougherara, H. A comprehensive review on surface modification of UHMWPE fiber and interfacial properties. *Compos. Part A Appl. Sci. Manuf.* **2021**, *140*, 106146. [CrossRef]
15. Li, W.W.; Li, R.P.; Li, C.Y.; Chen, Z.R.; Zhang, L. Mechanical Properties of Surface-Modified Ultra-High Molecular Weight Polyethylene Fiber Reinforced Natural Rubber Composites. *Polym. Compos.* **2017**, *38*, 1215–1220. [CrossRef]
16. Wang, L.; Gao, S.B.; Wang, J.J.; Wang, W.C.; Zhang, L.Q.; Tian, M. Surface modification of UHMWPE fibers by ozone treatment and UV grafting for adhesion improvement. *J. Adhes.* **2018**, *94*, 30–45. [CrossRef]
17. Fang, Z.H.; Tu, Q.Z.; Shen, X.M.; Yang, X.; Liang, K.; Pan, M.; Chen, Z.Y. Biomimetic surface modification of UHMWPE fibers to enhance interfacial adhesion with rubber matrix via constructing polydopamine functionalization platform and then depositing zinc oxide nanoparticles. *Surf. Interfaces* **2022**, *29*, 101728. [CrossRef]
18. Fang, Z.H.; Tu, Q.Z.; Chen, Z.Y.; Shen, X.M.; Pan, M.; Liang, K.; Yang, X. Study on catechol/tetraethylenepentamine and nano zinc oxide co-modifying ultrahigh molecular weight polyethylene fiber surface to improve interfacial adhesion. *Polym. Adv. Technol.* **2022**, *33*, 4072–4083. [CrossRef]
19. Li, W.W.; Meng, L.; Ma, R.L. Effect of surface treatment with potassium permanganate on ultra-high molecular weight polyethylene fiber reinforced natural rubber composites. *Polym. Test.* **2016**, *55*, 10–16. [CrossRef]
20. Sa, R.N.; Wei, Z.H.; Yan, Y.; Wang, L.; Wang, W.C.; Zhang, L.Q.; Ning, N.Y.; Tian, M. Catechol and epoxy functionalized ultrahigh molecular weight polyethylene (UHMWPE) fibers with improved surface activity and interfacial adhesion. *Compos. Sci. Technol.* **2015**, *113*, 54–62. [CrossRef]

21. Andideh, M.; Ghoreishy, M.H.R.; Soltani, S.; Sourki, F.A. Surface modification of oxidized carbon fibers by grafting bis (triethoxysilylpropyl) tetrasulfide (TESPT) and rubber sizing agent: Application to short carbon fibers/SBR composites. *Compos. Part A Appl. Sci. Manuf.* **2021**, *141*, 106201. [CrossRef]
22. Zhong, F.; Schwabe, J.; Hofmann, D.; Meier, J.; Thomann, R.; Enders, M.; Mulhaupt, R. All-polyethylene composites reinforced via extended-chain UHMWPE nanostructure formation during melt processing. *Polymer* **2018**, *140*, 107–116. [CrossRef]
23. Roy, K.; Debnath, S.C.; Das, A.; Heinrich, G.; Potiyaraj, P. Exploring the synergistic effect of short jute fiber and nanoclay on the mechanical, dynamic mechanical and thermal properties of natural rubber composites. *Polym. Test.* **2018**, *67*, 487–493. [CrossRef]
24. Wang, D.L.; Chen, S.; Chen, L.; Chen, B.Y.; Ren, F.J.; Zhu, C.X.; Feng, J. Investigation and improvement of the scorch behavior of silica-filled solution styrene-butadiene rubber compound. *J. Appl. Polym. Sci.* **2019**, *136*, 47918. [CrossRef]
25. Wang, D.L.; Ren, F.J.; Zhu, C.X.; Feng, J.; Chen, S.; Shen, G.L.; Wang, F.F. Hybrid silane technology in silica-reinforced tread compound. *Rubber Chem. Technol.* **2019**, *92*, 310–325. [CrossRef]
26. Meng, L.; Li, W.W.; Ma, R.L.; Huang, M.M.; Cao, Y.B.; Wang, J.W. Mechanical properties of rigid polyurethane composites reinforced with surface treated ultrahigh molecular weight polyethylene fibers. *Polym. Adv. Technol.* **2018**, *29*, 843–851. [CrossRef]

Disclaimer/Publisher's Note: The statements, opinions and data contained in all publications are solely those of the individual author(s) and contributor(s) and not of MDPI and/or the editor(s). MDPI and/or the editor(s) disclaim responsibility for any injury to people or property resulting from any ideas, methods, instructions or products referred to in the content.

Article

Effect of Different Silane Coupling Agents In-Situ Modified Sepiolite on the Structure and Properties of Natural Rubber Composites Prepared by Latex Compounding Method

Zhanfeng Hou, Dawei Zhou, Qi Chen * and Zhenxiang Xin

Key Laboratory of Rubber-Plastics, Ministry of Education, Shandong Provincial Key Laboratory of Rubber-Plastics, School of Polymer Science and Engineering, Qingdao University of Science and Technology, Qingdao 266042, China
* Correspondence: 03293@qust.edu.cn

Citation: Hou, Z.; Zhou, D.; Chen, Q.; Xin, Z. Effect of Different Silane Coupling Agents In-Situ Modified Sepiolite on the Structure and Properties of Natural Rubber Composites Prepared by Latex Compounding Method. *Polymers* 2023, 15, 1620. https://doi.org/10.3390/polym15071620

Academic Editors: Yuwei Chen and Yumin Xia

Received: 28 February 2023
Revised: 17 March 2023
Accepted: 19 March 2023
Published: 24 March 2023

Copyright: © 2023 by the authors. Licensee MDPI, Basel, Switzerland. This article is an open access article distributed under the terms and conditions of the Creative Commons Attribution (CC BY) license (https://creativecommons.org/licenses/by/4.0/).

Abstract: With the increasing demand for eco-friendly, non-petroleum-based natural rubber (NR) products, sepiolite, a naturally abundant, one-dimensional clay mineral, has been identified as a suitable material for reinforcing NR through the latex compounding method. To create superior NR/sepiolite composites, three silane coupling agents with different functional groups were used to modify sepiolite in situ via grafting or adsorption during the disaggregation and activation of natural sepiolite, which were subsequently mixed with natural rubber latex (NRL) to prepare the composites. The results showed that the modified sepiolite improved the dispersion and interfacial bonding strength with the rubber matrix. VTES-modified sepiolite containing C=C groups slightly improved the performance but retarded the vulcanization of the NR composites, and MPTES and TESPT-modified sepiolites containing -SH and $-S_4-$ groups, respectively, effectively accelerated vulcanization, inducing the composites to form a denser crosslink network structure, and exhibiting excellent dynamic and static properties, such as the modulus at a 300% increase from 8.82 MPa to 16.87 MPa, a tear strength increase from 49.6 N·mm^{-1} to 60.3 N·mm^{-1}, as well as an improved rolling resistance and abrasive resistance of the composites. These findings demonstrate that modified sepiolite can be used to produce high-quality NR/sepiolite composites with enhanced properties.

Keywords: natural rubber (NR); sepiolite; silane coupling agent; crosslink network structure; dynamic and static properties

1. Introduction

Natural rubber is a highly elastic material that is obtained from the natural latex of rubber trees through a series of processing steps [1]. It is widely used in various fields such as tire production, transmission, and transportation [2,3]. To enhance its performance and reduce costs, reinforcing fillers are added to natural rubber. Carbon black [4,5] and silica [6,7] are the most commonly used reinforcing materials in the rubber industry. However, the production of carbon black relies on non-renewable petroleum-based energy sources, while silica production consumes a significant amount of energy and causes environmental pollution [8]. To address these challenges and promote sustainability, researchers have sought alternative materials, such as new structured carbon-based materials such as graphene [9] and carbon nanotubes [10,11], bio-based materials such as cellulose [12,13], and clay mineral materials [14–16] such as montmorillonite and kaolin. Clay mineral materials are of particular interest due to their abundant reserves, easy accessibility, and low cost. Moreover, clay minerals have diverse structural forms, such as two-dimensional lamellar montmorillonite [17] and kaolinite, as well as one-dimensional fibrous sepiolite and palygorskite [18,19], which can achieve high levels of reinforcement after appropriate treatment. Therefore, they have become a popular choice for enhancing the performance of natural rubber.

Sepiolite is a naturally abundant, non-toxic, one-dimensional fibrous silicate material that has an ideal structural formula of $Si_{12}O_{30}Mg_8(OH)_4(H_2O)_4 \cdot 8H_2O$ [20], with fiber lengths ranging from 0.2 μm to 5 μm [21]. Its unique structure comprises two continuous tetrahedral sheets and one discontinuous octahedral sheet, which create many tunnels and channels, leading to a large specific surface area. Sepiolite also has a high density of silanol groups, which form as a result of the combination between non-shared oxygen atoms of the tetrahedral silicon sheets and hydrogen [22]. Sepiolite's special fibrous crystal morphology, large specific surface area, and abundant silanol groups give it excellent adsorption, enhancement, and stable suspension in the aqueous phase. However, natural sepiolite fibers tend to exist more as aggregates or bundles due to van der Waals forces between fibers. Moreover, complex mineral formation conditions often lead to the co-existence of trace-associated minerals with sepiolite, reducing its specific surface area and surface activity [23,24]. These factors can limit its dispersion in the polymer matrix and limit its application as a nanomaterial.

To achieve excellent natural rubber/sepiolite composites, two key challenges in their processing and application must be addressed: the dispersibility of sepiolite within the rubber matrix and the strength of the interfacial bonding between sepiolite and natural rubber [25]. By enhancing the dispersion of sepiolite in the natural rubber matrix and leveraging the intrinsic properties of sepiolite and natural latex, these composites can be prepared using the economical and environmentally friendly latex compounding method [26,27]. This approach [28] offers several advantages over traditional melt mixing, including lower energy consumption, reduced dust pollution, and improved filler dispersion in the rubber matrix, resulting in an enhanced composite performance. In the latex compounding method, sepiolite must be disaggregated and activated to prepare homogeneous sepiolite dispersions. The most effective approach to achieve this involves ultrasonic disaggregation combined with acid-thermal activation, such as that of Ruiz-Hitzky [29,30], who successfully prepared a highly stable sepiolite suspension system using ultrasonic means, while Jiménez-López [31] and Zhou et al. [23] employed thermal activation by HNO_3 and microwave-assisted thermal activation by HCl, respectively, leading to a more significant increase in the specific surface area and surface activity of sepiolite. To improve the strength of the sepiolite–polymer interface, researchers have used various approaches. Hayeemasae [32] utilized sepiolite-reinforced epoxidized natural rubber, leveraging strong interactions between sepiolite's hydroxyl and siloxane groups and epoxy groups. Raji [33] and Peinado [34] modified sepiolite with aminosilanes and added it to polypropylene and poly(lactic acid), respectively, to enhance the compatibility of sepiolite with polymers and material properties. Silane coupling agent-modified sepiolite, as used by Wang et al. [35] to reinforce cis-polybutadiene rubber, significantly improved the mechanical properties of the resulting composites, particularly when KH560 was used at 7%, which increased tensile and tear strengths by 108.3% and 74.1%, respectively. These results suggest that the addition of a silane coupling agent has a substantial impact on improving the strength of the sepiolite–polymer interface.

Previous research [32,36,37] has primarily focused on melt mixing, and there is limited information on in situ modification of sepiolite for latex compounding. In this study, we selected three silane coupling agents with different functional groups (see Table 1) to modify sepiolite. Our modification mechanism [38–40] involved the hydrolysis of $Si–O–C_2H_5$ in the silane coupling agent and the subsequent condensation of hydroxyl groups on the surface of sepiolite, with C=C in VTES, -HS- in MPTES, and $–S_4–$ in TESPT all able to participate in the vulcanization process of natural rubber and form strong chemical bonds. Hydrolysis of silane coupling agents is known to be slow, often requiring the addition of acid to promote hydrolysis [41,42] and improve the efficiency of hydroxyl condensation with the inorganic filler surface. Capitalizing on the acidic conditions of sepiolite during depolymerization activation, we employed a one-step activation modification method to prepare in situ silane-modified sepiolite. This method not only improved the efficiency of sepiolite modification but also reduced energy and acid consumption. Morphology, activity, and modification levels of the modified sepiolite were characterized using X-ray diffraction (XRD), Fourier-transform infrared spectroscopy (FTIR), scanning electron microscopy (SEM), and thermogravimetric analysis (TGA). Next, we prepared natural rubber/sepiolite composites by emulsion mixing of the in situ modified sepiolite with natural latex, evaluating the dispersibility and interfacial binding ability using tensile section morphology, DSC, and bound rubber content. The effects of modified sepiolite on the vulcanization characteristics and the dynamic and static properties of the composites were also analyzed.

Table 1. Structure and chemical characteristics of silane molecules.

Chemical Name	Functional Group	Structural Formula	M (g/mol)
Triethoxyvinylsilane (VTES)	vinyl		190.31
3-Mercaptopropyltriethoxysilane (MPTES)	mercapto		238.42
Bis[3-(triethoxysilyl)propyl] tetrasulfide (TESPT)	tetrasulfide		538.95

2. Experimental Materials and Methods

2.1. Materials

A thirty-six percent total solid content of low-ammonia NRL was obtained from the Chinese Academy of Tropical Agricultural Sciences (Danzhou, China). Sepiolite concentrate (X-ray fluorescence (XRF) chemical analysis indicated that the composition of this fraction is 65.19% SiO_2, 20.25% MgO, 7.89% Al_2O_3, 3.11% Fe_2O_3, 1.05% CaO) was obtained from Qingdao Zhongxiang Environmental Protection Technology Co., Ltd (Qingdao, China). The silanes, namely Triethoxyvinylsilane (VTES; 97%; M_w = 190.31 g/mol), (3-Mercaptopropyl)Triethoxysilane (MPTES; 98%; M_w = 238.42 g/mol), and Bis[3-(Triethoxysilyl)Propyl]Tetrasulfide (TESPT; 90%; M_w = 538.95 g/mol), were purchased from Shanghai Macklin Biochemical Technology Co., Ltd. (Shanghai, China). Oxalic acid dihydrate (OA, $C_2H_2O_4 \cdot 2H_2O$, AR), zinc oxide (ZnO), and stearic acid (SA) were purchased from Sinopharm Chemical Reagent Co., Ltd. (Shanghai, China). The N-cyclohexyl benzothiazole-2-sulphonamide (Accelerator CZ), 2,2′-dibenzothiazoledisulfde (Accelerator DM), and sulfur (S) were industrial grade and provided by SanLux Co., Ltd. (Shaoxing, China).

2.2. Preparation of Sepiolite Dispersions

To begin, 20 g of sepiolite powder was added to an aqueous solution at a solid-to-liquid ratio of 1:20 (w/w). The mixture was then sonicated using a TiAl-V tip sonicator (SCIENTZ JY99-IIDN, with a 22 mm diameter tip, Ningbo, China) in pulses of 5 s on and off for a total of 12 min to achieve a homogeneous suspension. Next, 6 g of oxalic acid was added to the sepiolite suspension, and the mixture was stirred at 80 °C for 6 h. The pH of the suspension was then adjusted to a range between 3.5 and 4.5 by rinsing with deionized water. After this, 2 g of VTES, MPTES, or TESPT silane coupling agents was added to the suspension and stirred for an additional 4 h at 80 °C. The suspension was then washed with deionized water until it reached a pH of 7, resulting in the in situ modified sepiolite with a silane coupling agent. The modified sepiolite samples were named VTES-Sep, MPTES-Sep, and TESPT-Sep, while a comparison sample was synthesized using only oxalic acid and neutral pH washing, and was named Sep.

To obtain part of the pure sepiolite and modified sepiolite powder, the corresponding modified sepiolite slurry was dried. The sepiolite powders were extracted using ethanol in a Soxhlet extractor for 24 h, with 15 min reflux intervals, to remove any un-grafted silane coupling agent. Finally, the extracted sepiolite powders were dried in an oven at 80 °C for 24 h and were prepared for characterization using XRD, FTIR, and TGA.

2.3. Preparation of Sep/NR Masterbatches and Composites

The NRL was diluted with deionized water to a concentration of 20 wt%. Sepiolite suspensions were then separately mixed with NRL by stirring at 500 rpm for 30 min, with a mass ratio of 20% sepiolite to NR (e.g., 20 g dry weight of sepiolite for every 100 g dry weight of NR). The mixture was then flocculated with a 2% $CaCl_2$ solution, washed with water, and dried in a vacuum oven at 60 °C to obtain sepiolite/NR masterbatches.

The masterbatches were plasticized eight times in a two-roll open mill. Then, the ingredients for vulcanization and other additives were added one-by-one to the masterbatch, with a total mixing time of 10 min. The compounds were then vulcanized using an XLR-D vulcanizer at 150 °C under a pressure of 10 MPa for the optimum cure time (t90), as determined using a non-rotor curemeter. After curing, the samples were air-cooled to obtain the composites. The composites are coded as Sep/NR, VSep/NR, MSep/NR, and TSep/NR. The process of sepiolite modification and latex compounding is shown in Figure 1, and the formulation of sepiolite/NR compounds is demonstrated in Table 2.

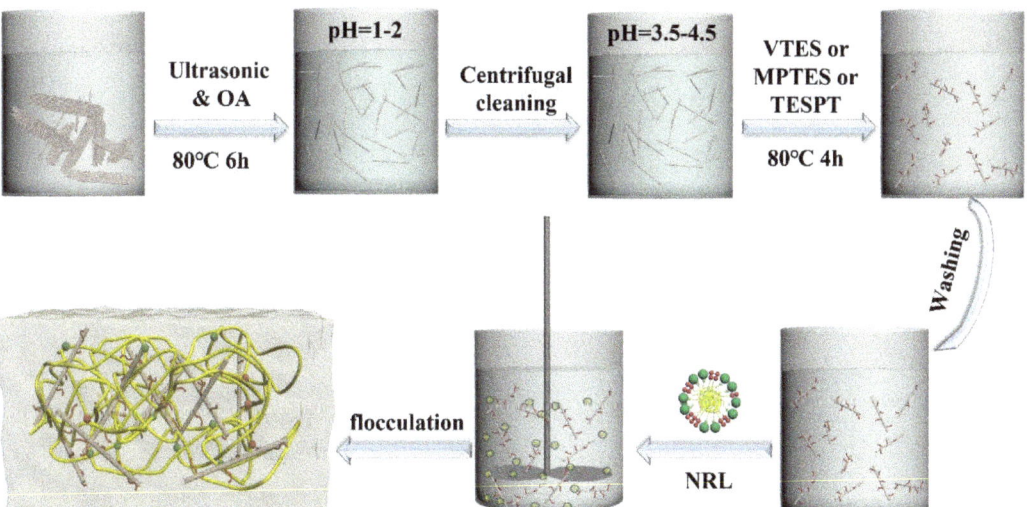

Figure 1. Schematic diagram of sepiolite modification and the latex compounding method.

2.4. Characterizations

XRD images of the sepiolites were obtained by a Rigaku D-MAX 2500-PC diffractometer (Tokyo, Japan) with nickel-filtered Cu Kα radiation of λ = 0.154 nm. The scanning rate was 5°/min, and the test angle was 5–70°.

Table 2. Formulation of sepiolite/NR composites, phr [a].

Materials	Sep/NR	VSep/NR	MSep/NR	TSep/NR
Masterbatches	120 [b]	120	120	120
VTES	0	2	0	0
MPTES	0	0	2	0
TESPT	0	0	0	2
zinc oxide	5	5	5	5
stearic acid	5	5	5	5
Accelerator CZ	2	2	2	2
Accelerator DM	1	1	1	1
sulfur	2	2	2	2

[a] Parts per hundred of rubber. [b] 120 phr of the masterbatch = 100 phr of NR + 20 phr of sepiolite.

FTIR spectra of sepiolite were recorded on a Bruker VERTEX 70 spectrometer (Bruker Optik GmbH Co., Ettlingen, Germany) by averaging 32 scans at a 4 cm^{-1} resolution, with the wavenumber ranging from 4000 to 400 cm^{-1}.

TGA was used to evaluate the thermal degradation of sepiolite and the impact of silane functionalization on sepiolite. The analysis was carried out using a STA449(F5) Thermogravimetric Analyzer (NETZSCH-Gerätebau GmbH, Selb, Germany). To accurately determine the number of silane graft modifications, a representative sample was heated in a platinum pan under air from room temperature to 800 °C, with a heating rate of 10 °C/min. The amount of grafted and intercalated silane molecules has been calculated using the following equation [33,43]:

$$\text{grafted amount (mequiv/g)} = \frac{10^3 W_{150-650}}{(100 - W_{150-650})M} \quad (1)$$

where $W_{150-650}$ is the number of silane degradation between 150 and 650 °C, and M (g/mol) is the molecular weight of the grafted silane molecules.

The surface morphology of sepiolites and tensile fractured surfaces of vulcanizates were observed by SEM performed on a JSM-7500F (JSOL, Tokyo, Japan). All specimens were sputtered with gold before observations.

The bound rubber content [25] was measured on un-vulcanized compounds. Firstly, 0.5 g of the un-vulcanized compounds were cut into small pieces and put into a sample cage prepared by nickel mesh (400 mesh). Then, the sample cage was placed in a frosted glass bottle with toluene and immersed for 72 h at room temperature. The toluene was replaced every 24 h. Lastly, the residual sample was taken out from the toluene and dried at 80 °C in a vacuum to a constant weight. Three samples of each group were tested, and the average was taken as the final result.

The bound rubber content was calculated according to the following equation:

$$\text{Bound rubber content} = \frac{M_1 - M_0 \times f}{M_0 - M_0 \times f} \quad (2)$$

where M_0 is the initial weight of the sample, M_1 is the mass of the sample dried to constant weight, and f is the weight fraction of sepiolite in the compound.

The specific heat capacity curves were acquired through the differential scanning calorimeter test (DSC, NETZSCH-204, NETZSCH, Selb, Germany). The samples were performed at a 10 °C/min heating rate at −100–25 °C in a nitrogen atmosphere. The normalized specific heat capacity step (ΔC_{pn}) and the mass fraction of the immobilized polymer layer (χ_{im}) were calculated as follows [44,45]:

$$\Delta C_{pn} = \Delta C_p / (1 - w) \quad (3)$$

$$\chi_{im} = \frac{\Delta C_{p0} - \Delta C_{pn}}{\Delta C_{p0}} \quad (4)$$

where ΔC_p is the heat capacity jump at T_g and was obtained by the software NETZSCH Thermal Analysis, ω is the weight fraction of the filler, and ΔC_{p0} indicates the specific heat capacity variation at T_g of unfilled NR.

Curing characteristics were evaluated using a rotorless rheometer (MDR, Alpha Technologies, Akron, OH, USA) at 150 °C, and the Flory–Rehner equation [46] was used to determine the crosslinking density based on the equilibrium swelling method with toluene as the solvent. Toluene has a solubility parameter (18.2) similar to that of natural rubber (16.2–17.0). Three measurements were conducted for each sample, and the mean values with statistical errors are presented.

Tensile tests, stress relaxation experiments, and tear tests were conducted on a Zwick Roell material testing machine (Z005, Zwick/Roell GmbH Co., Ulm, Germany). Type 2 dumbbell samples prepared according to ISO37-2005 [47] were used to perform the stress relaxation and stress–strain tests at 25 °C. Stress–strain curves were obtained by carrying out simple uniaxial tension tests at an extension rate of 500 mm/min. The stress relaxation curves were recorded at a constant strain of 100% for 1000 s. Tear tests were performed on angle test pieces (approximately $100 \times 20 \times 2$ mm^3) at an extension rate of 500 mm/min, following the ISO 34-1-2015 standard [48]. Five measurements were conducted for each sample, and the average value with statistical errors is reported.

The strain-dependent storage modulus (G′) and the loss factor (tanδ) of the rubber compounds were analyzed using a rubber process analyzer (RPA2000, Alpha Technologies, Akron, OH, USA). The strain amplitude changed from 0.28% to 100% at the test frequency of 1 Hz and a temperature of 60 °C.

The dynamic mechanical analysis (DMA) was carried out in a tension mode on a Dynamic Thermomechanical Analyzer (DMTS, EPlexor 500N, NETZSCH-Gerätebau GmbH, Selb, Germany). The dumbbell samples of type 2 with dimensions of ca. $75 \times 4 \times 2$ mm^3 (ISO-37-2005) [47] and a test length of 10 mm were cut from the vulcanizate sheets. The measurements were performed at temperatures between -80 °C and 80 °C with a heating rate of 3 °C/min at a dynamic strain of 0.1%, a static strain of 0.5%, and a frequency of 10 Hz.

Abrasive resistance was evaluated by a DIN abrader (GT-7012D, GOTECH Testing machines Co., Ltd., Taiwan, China) with a standard of ISO 4649-2017 [49]. The reported values were averaged from three independent results of volume loss.

3. Results and Discussion

3.1. Characterization of Pure and Modified Sepiolite

XRD analysis was utilized to investigate any changes in the crystal structure of sepiolite. Figure 2A illustrates the XRD patterns of sepiolite and silane-modified sepiolite. Upon comparison with the standard JCPDS map for sepiolite, it was found that all reflections of unmodified sepiolite were consistent with it, and no additional diffraction peaks were detected. This indicates that the purity of sepiolite was improved after undergoing disaggregation activation treatment. The XRD patterns of the silane-modified sepiolite were similar to those of unmodified sepiolite, with characteristic reflections at 2θ = 7.3° (d = 12.1 Å), 20.6° (d = 4.31 Å), and 35.0° (d = 2.56 Å). This observation confirms that the silane modification does not alter the crystal structure of sepiolite, consistent with prior findings by Tartaglione et al. [50].

In Figure 2B, the FTIR spectra of sepiolite and silane-modified sepiolite are shown. Sepiolite contains various types of water molecules, including adsorbed water, zeolitic water, bound water, and structural water, which are present inside the channels or on the surface [51]. The stretching vibrations of (Mg/Al)−OH groups and Si−OH groups of sepiolite were assigned to absorption bands at 3626 cm^{-1} and 3526 cm^{-1}, respectively [52]. The stretching vibration of −OH, primarily from surface-adsorbed water and zeolitic water, resulted in a broad band centered at 3420 cm^{-1}. The H−O−H bending vibration of zeolitic water and bound water led to the appearance of a band at 1660 cm^{-1}. The antisymmetric stretching vibration and stretching vibration of the Si−O−Si group of the tetrahedral sheets caused the bands at 1204 cm^{-1} and 1026 cm^{-1}, respectively [51]. Additionally, the appearance of new bands at 2929 and 2850 cm^{-1} in VTES−Sep, MPTS−Sep, and TESPT−Sep, corresponding to the antisymmetric and symmetric stretching vibration of C−H in organosilanes, respectively, indicates the successful grafting of VTES, MPTS, or TESPT onto sepiolite [53].

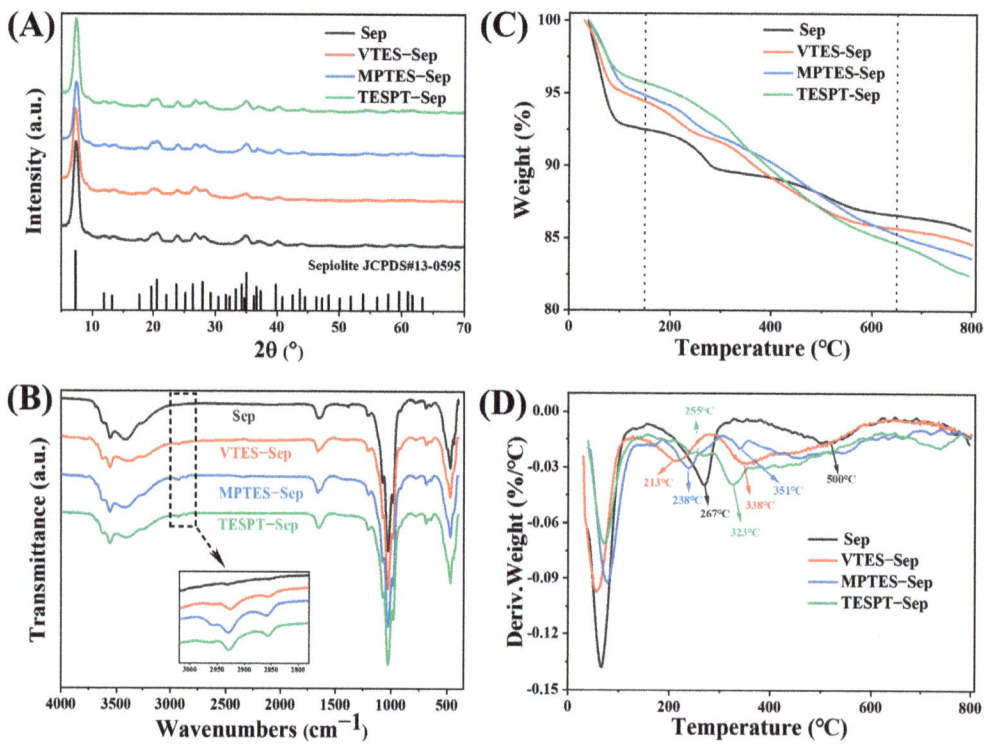

Figure 2. Characteristics of pure and modified sepiolite. (**A**) XRD patterns, (**B**) FTIR spectra, (**C**) TGA curves, and (**D**) derivative TGA curves.

The thermogravimetric (TG) and derivative thermogravimetric (DTG) curves of sepiolite and silane-modified sepiolites are presented in Figure 2C,D, and their weight losses are summarized in Table 3. The thermogravimetric curves of unmodified sepiolite displayed four discrete weight losses. The initial weight loss before 150 °C was attributed to the evaporation of adsorbed water and zeolite water [54] (i.e., adsorbed on the external surface and in the structural channels), with a loss of 7.5%. Although, this value is not entirely consistent with the literature [54,55] and is mainly related to the environmental humidity and the hydrophilicity of sepiolite. The elimination of bound water occurred in two stages [56], from 150 °C to 400 °C and 400 °C to 650 °C, with maximum weight loss temperatures of 267 °C and 500 °C, respectively, resulting in a total weight loss of 5.9%. As the temperature increased, the hydroxyl groups in the sepiolite condensed and dehydrated, ultimately leading to complete structural damage, with a loss of 1.02% between 650 °C and 800 °C. The weight loss of silane-modified sepiolite was reduced up to 150 °C, indicating that the hydrophilicity of the modified sepiolite was reduced. DTG analysis in Figure 2D revealed that the volatilization of the modifier on modified sepiolite was divided into two distinct steps. The first step occurred at a relatively low temperature (T < 267 °C), and the weight loss of VTES−Sep and MPTS−Sep was more pronounced. This result is attributed to sepiolite having excellent adsorption properties, where due to its porous structure and the abundance of silanol groups on its surface, hydrolyzed silane molecules are easily adsorbed on sepiolite through hydrogen bonding or van der Waals forces [50]. The second step occurred at a relatively high temperature and was dominated by the volatilization of the grafted modifier [33]. The amount of silane modification was calculated based on the volatilization mass between 150 °C and 650 °C, indicating that the percentage of silane molecule grafted on sepiolite was about 2.83% for VTES-Sep, 3.66% for MPTS−Sep, and 5.16% for TESPT−Sep. Furthermore, the number of intercalated molecules that effectively participated in the silylation reaction was estimated to be 0.282 mequiv/g for VTES-Sep, 0.251 mequiv/g for MPTS-Sep, and 0.120 mequiv/g for TESPT−Sep. These results provide compelling evidence for the successful silylation of sepiolite.

Table 3. Thermogravimetric analysis values of sepiolite and silane-modified sepiolites.

Materials	Weight Loss/%			Modifier/%	Grafted Amount/(mequiv/g)
	40–150 °C	150–650 °C [a]	650–800 °C		
Sep	7.50	5.90	1.02	-	-
VTES-Sep	5.57	8.73	1.06	2.83	0.282
MPTES-Sep	5.13	9.56	1.66	3.66	0.251
TESPT-Sep	4.27	11.06	2.28	5.16	0.120

[a] Silane modifier was evaluated between 150 °C and 650 °C.

SEM images (Figure 3) were used to investigate the microstructure and morphological changes of sepiolite, with the sample being prepared by adding a diluted suspension of sepiolite to the sample table, followed by drying and gold sputtering. The SEM images revealed that the shape structure and aggregation morphology of sepiolite before and after modification remained largely unchanged, with rod and micro-bundle shapes being observed in all cases (Figure 3a–d). At a magnification of 10,000 times (Figure 3a′–d′), it was observed that the surface of unmodified sepiolite was relatively smooth, whereas the surfaces of sepiolite modified by silane coupling agents displayed distinct changes. Specifically, sepiolite modified by VTES exhibited a greater number of spherical protrusions, similar in shape to those observed in sepiolite modified by VTMS [57], and the surfaces of sepiolite modified by MTPS and TESPT showed more coverage and roughness. These observations indicate that all three silane coupling agents successfully graft onto the surface of sepiolite.

3.2. Dispersion and Interfacial Interaction of Sepiolite/NR Composites

In order to visually characterize the dispersion and interfacial interactions of sepiolite in natural rubber, we employed SEM to observe the morphology of the tensile fracture surface of vulcanized rubber. Figure 4a,a′ reveals that the fracture surface of the Sep/NR vulcanizates is relatively smooth, with some sepiolite rod-like particles visibly exposed on the surface, and a small amount of particles aggregated. This suggests that while the wet compounding process can improve sepiolite dispersion in the matrix, unmodified sepiolite exhibits lower compatibility with the natural rubber matrix. In contrast, Figure 4b–d,b′–d′ depict the cross-sections of VSep/NR, MSep/NR, and TSep/NR vulcanizates, respectively. It can be seen that sepiolite particles are distributed in the natural rubber matrix as individual rods without any apparent aggregation, and most of the rod-shaped particles are embedded in the rubber matrix. Especially in MSep/NR and TSep/NR composites, sepiolite almost fused with the natural rubber, resulting in a blurred interface. These observations suggest that silane-modified sepiolite has better compatibility with the natural rubber matrix and can be more effectively dispersed within the rubber matrix, which improves the interfacial bond strength with the matrix, with MPTS and TESPT modifications demonstrating greater application efficacy than VTES modification.

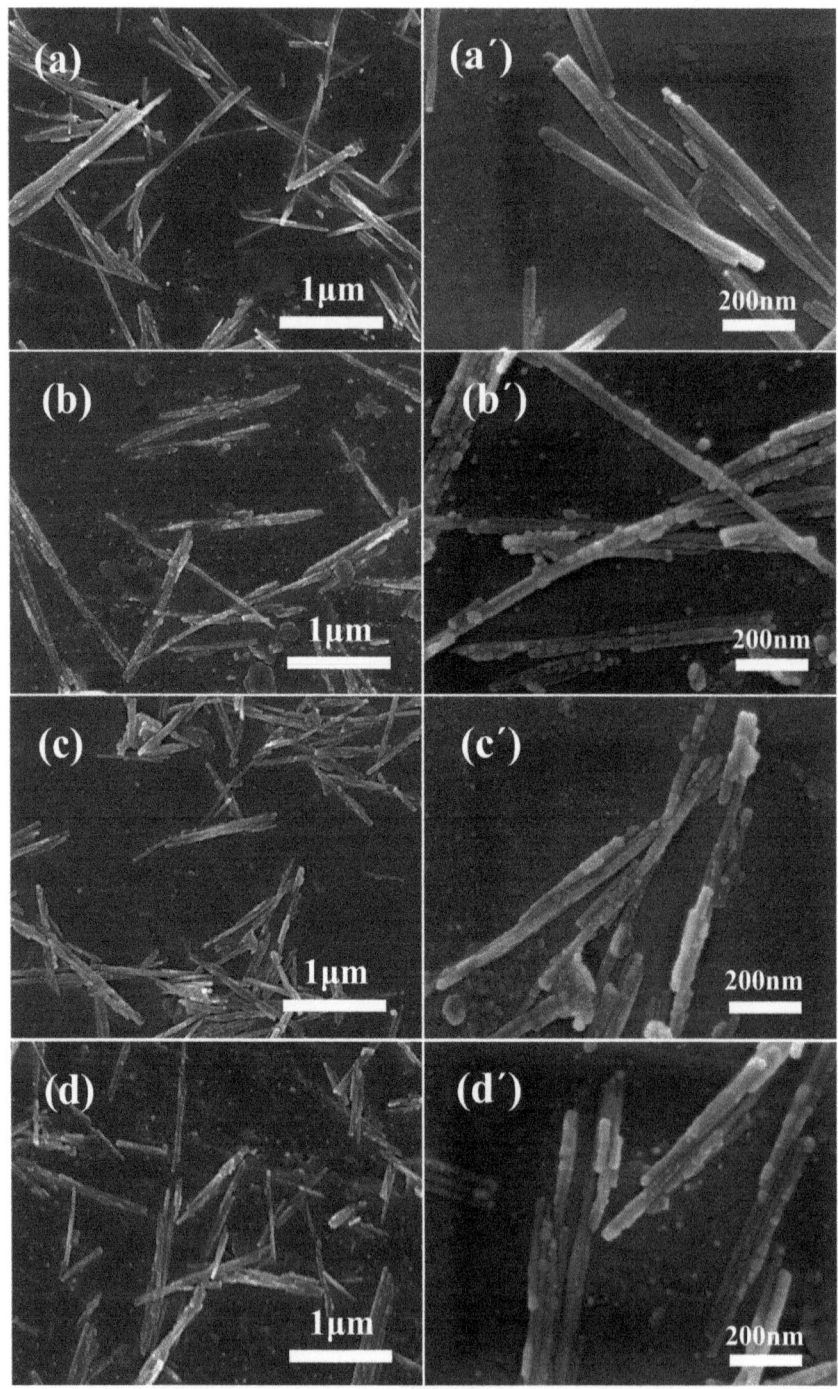

Figure 3. Microscopic morphology of sepiolite and silane-modified sepiolites. (**a,a′**) Sep, (**b,b′**) VTES−Sep, (**c,c′**) MPTES−Sep, and (**d,d′**) TESPT−Sep (left ×20,000, right ×100,000).

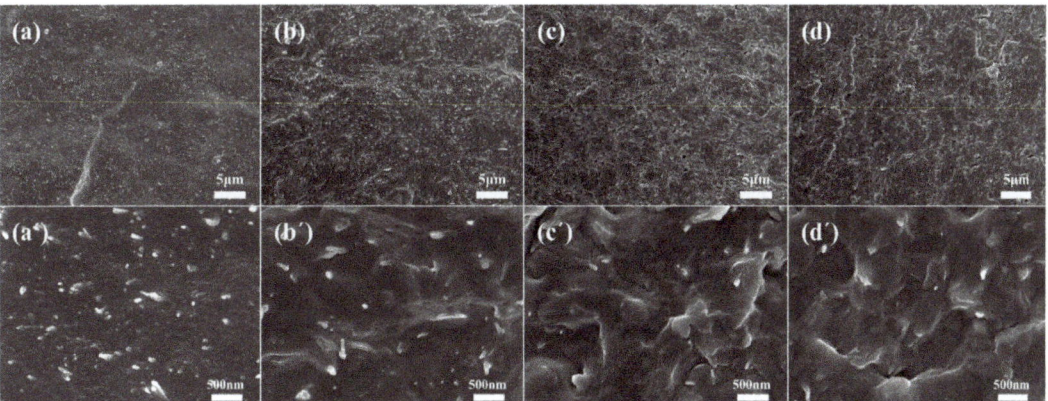

Figure 4. SEM micrographs of tensile fractured surfaces of (**a,a′**) Sep/NR composite, (**b,b′**) VSep/NR composite, (**c,c′**) MSep/NR composite, and (**d,d′**) TSep/NR composite.

The storage modulus (G′) of uncured composites was measured as a function of strain amplitude and is shown in Figure 5. At low strains, all samples exhibited a rapid decrease in G′ with increasing strain amplitude, resulting in non-linear viscoelastic behavior known as the Payne effect [58]. This effect can be used to assess the filler network of the composites based on the difference between the maximum and minimum G′ (ΔG′). The Sep/NR sample exhibited the strongest Payne effect, indicating a strong filler network and poor dispersion. The ΔG′ of the composites with silane-modified sepiolite showed a significant decrease, indicating an improvement in its dispersion in the rubber matrix and a reduction in the formation of its own filler network.

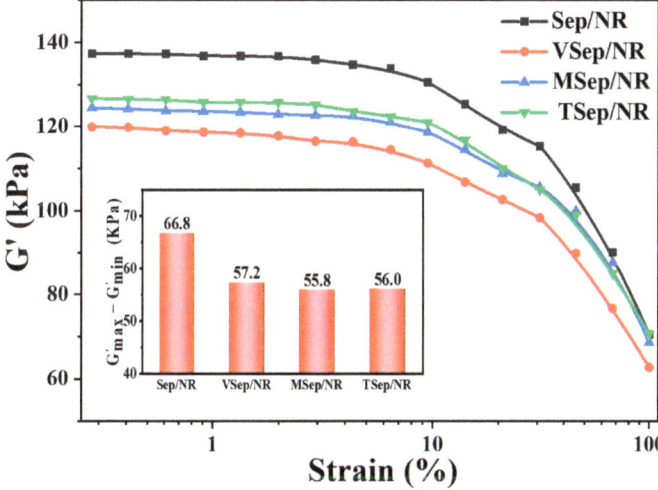

Figure 5. Strain amplitude dependence of the storage modulus (G′) of sepiolite/NR composites.

The bound rubber content, which is defined as the indissoluble rubber in good solvents, can be used to characterize the interaction between the rubber and filler. The higher the bound rubber content, the stronger the interaction. Figure 6 displays the bound rubber content of all samples. The Sep/NR composites had the lowest bound rubber content at 21.3%, which was mainly due to the adsorption or entanglement of rubber molecular chains with the sepiolite. The composites with silane-modified sepiolite exhibited a significantly higher bound rubber content, with TSep/NR and MSep/NR similar at 32.1% and 31.6%, respectively, and VSep/NR at 24.9%. This is because the silane coupling agent is adsorbed and grafted onto the surface of sepiolite, which can entangle with more rubber molecular chains. Additionally, during the subsequent mixing process, due to the high local temperature, $-S_4-$ and -SH in TESPT−Sep and MPTES−Sep were activated and combined with rubber molecular chains to form chemical bonds, producing tightly bound rubber [59]. While C=C in VTES relies only on external sulfur addition to produce chemical bonding, the amount of bonded rubber formed is less due to the lower temperature of the applied sulfur.

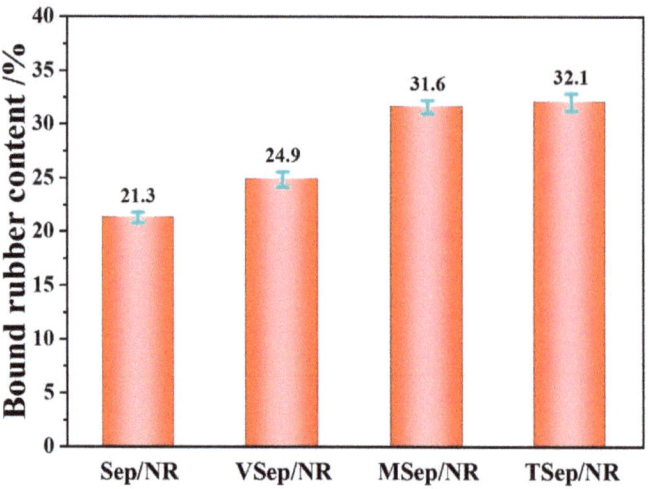

Figure 6. The bound rubber contents of sepiolite/NR composites.

The interfacial interaction between rubber and sepiolite can also be characterized by the mobility of rubber chain segments at and near the sepiolite particle surface. DSC was used to analyze the ΔC_p of sepiolite/NR composites, as shown in Figure 7A, and the normalized specific heat capacity step (ΔC_{pn}) and mass fraction of the immobilized polymer layer (χ_{im}) of sepiolite/NR composites are illustrated in Figure 7B. The values of ΔC_p for NR/sepiolite composites were all lower than the Neat NR (the natural rubber contains zinc oxide, stearic acid, CZ, DM, and sulfur in the same amount as in Table 2, but without other reinforcing materials such as sepiolite), indicating restricted movement of molecular chains and chain segments due to the addition of sepiolite. Comparing the samples with silane coupling agents to Sep/NR samples, the ΔC_p and ΔC_{pn} values of the modified blends were significantly lower, and the mass fraction of the immobilized polymer layer (χ_{im}) was significantly increased. This indicated tighter bonding between natural rubber and sepiolite and improved interfacial strength. Comparing VSep/NR, MSep/NR, and TSep/NR, it was observed that the χ_{im} value of TSep/NR was the largest, followed by MSep/NR and VSep/NR, suggesting that there is a difference in the bonding ability between sepiolite and the rubber matrix after modification with different coupling agents. TESPT-modified sepiolite had the strongest bonding ability between sepiolite and the natural rubber matrix, and the interfacial strength was the largest, which was consistent with the change in the bound rubber content to the composite.

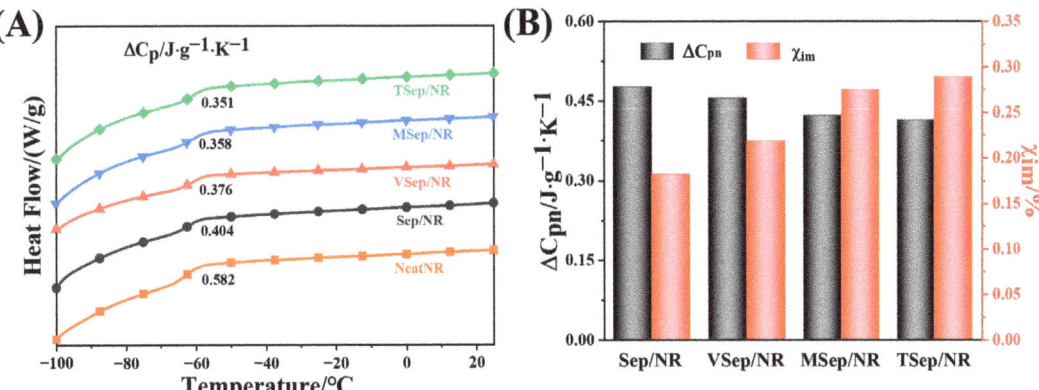

Figure 7. (**A**) DSC curves and (**B**) ΔC_{pn} and χim, of sepiolite/NR composites.

3.3. Vulcanization Characteristics of Sepiolite/NR Composites

Curing is a crucial process in the production of rubber products. Figure 8A and Table 4 illustrate the curing characteristics of sepiolite/natural rubber composites. The scorch time (t10) represents the degree of early vulcanization. By increasing the scorch time, the occurrence of early crosslinking in linear molecules within the compound is reduced, leading to a lower likelihood of premature vulcanization. The optimum vulcanization time (t90) of the compound is shortened as the crosslinking speed of linear molecules is accelerated, resulting in a faster attainment of the maximum crosslinking density [60]. It is observed that the t10 and t90 of VSep/NR have increased compared to Sep/NR. This is because the functional group of VTES contains a C=C double bond, which increases the number of double bonds in the VTES-Sep and the natural rubber blend. As a result, the time required to achieve equilibrium crosslinking is prolonged. On the other hand, the t10 and t90 of MSep/NR and TSep/NR have decreased, indicating accelerated vulcanization rates. This is because the functional groups in MPTES and TESPT both contain −S−, which can participate in crosslinking reactions. The higher reactivity of -SH in MPTES resulted in a significantly faster vulcanization time of the composites. The minimum torque (M_L) is related to the dispersion of the filler and the network structure in the compounded rubber. Table 4 shows that the M_L of all composites decreased compared to Sep/NR, suggesting that the silane modification of sepiolite improved its dispersion and restricted its network structure. The difference between the maximum torque and the minimum torque (M_H-M_L) represents the stiffness of the rubber composite, which is positively associated with the crosslinking density and the interaction between the filler and rubber. Table 4 and Figure 8B show that the M_H-M_L and crosslinking density of the composites with modified sepiolite increased compared to Sep/NR. For VSep/NR composites, which rely solely on the higher dispersion of sepiolite, a slight increase in crosslinking density was shown [61], while for MSep/NR and TSep/NR, the presence of −S− groups in modified sepiolite can enhance the crosslinking density, where TESPT−Sep has the largest amount of grafting and the largest amount of −S−, resulting in the highest M_H-M_L value and the highest crosslinking density [62].

Table 4. The curing characteristics of sepiolite/NR composites.

Samples	t10/min	t90/min	M_L/dN·m	M_H/dN·m	M_H-M_L/dN·m
Sep/NR	5.29	8.89	0.84	10.09	9.25
VSep/NR	5.37	11.19	0.57	10.21	9.64
MSep/NR	3.33	6.94	0.57	10.55	9.98
TSep/NR	4.91	8.39	0.65	10.81	10.16

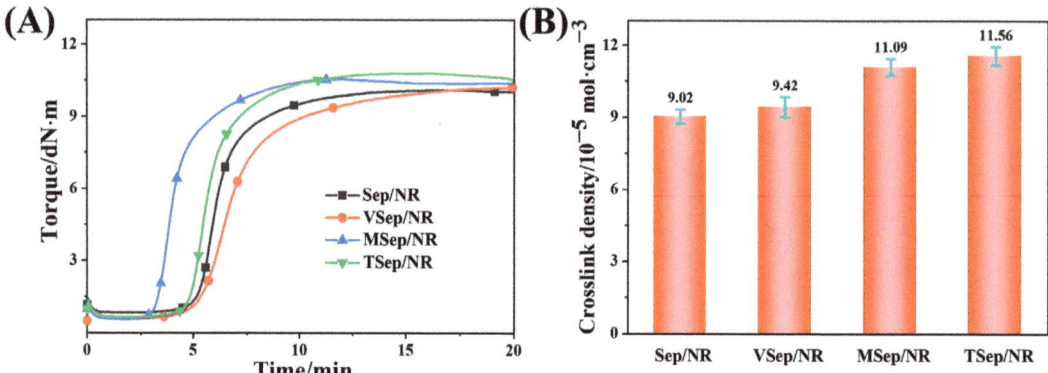

Figure 8. Characteristics of sepiolite/NR composites: (**A**) the curing curves and the (**B**) crosslinking density.

3.4. Static Mechanical Properties of Sepiolite/NR Composites

Figure 9A illustrates the stress–strain behavior of sepiolite/NR composites, while Table 5 presents the corresponding static mechanical performance data. The static mechanical properties of VSep/NR were observed to have slightly improved compared to Sep/NR, which can be attributed to the improved dispersion of sepiolite. On the other hand, the tensile strength and elongation at break of MSep/NR and TSep/NR composites were found to decrease, but the modulus and tear strength were significantly enhanced. Specifically, the modulus at 300% of MSep/NR and TSep/NR increased from 8.82 MPa to 14.99 MPa and 16.87 MPa, respectively, representing an increase of 70% and 91%. Moreover, the tear strength increased from 49.6 N·mm^{-1} to 58.1 N·mm^{-1} and 60.3 N·mm^{-1}, corresponding to a percentage increase of 17.1% and 21.6%, respectively. A higher modulus and tear strength are critical for certain dynamic applications of rubber products.

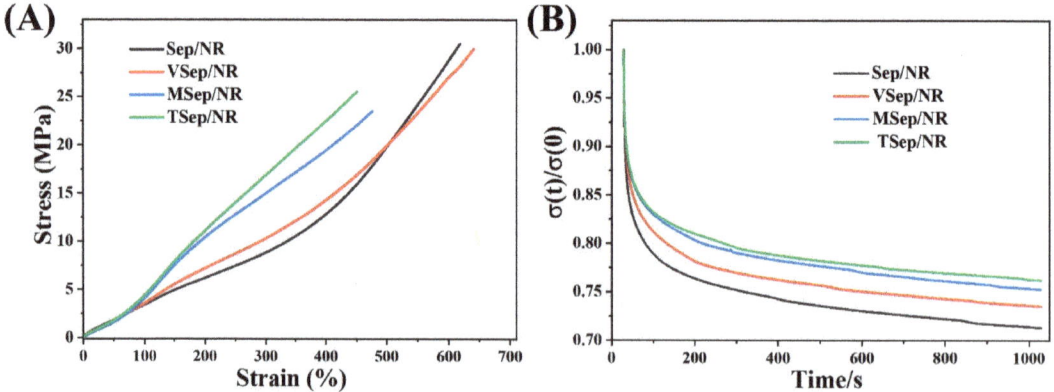

Figure 9. (**A**) The stress–strain curves of sepiolite/NR vulcanized composites, and (**B**) stress relaxation curves normalized with respect to the initial stress for vulcanizates (applied strain 100%).

To gain a deeper understanding of the different network structures in the nanocomposites, the relaxation behavior of the nanocomposites under stress was characterized, as shown in Figure 9B. The stress relaxation curves of the four types of vulcanized rubbers under 100% strain for 1000 s were normalized by their respective initial stresses. All four vulcanizates displayed a typical stress relaxation behavior, where the stress rapidly decreased at the beginning and then decreased as the system approached equilibrium [63]. The stress relaxation of the Sep/NR composite was the largest, with the lowest stress value at equilibrium, owing to its low bound rubber content and fewer crosslinking networks. In contrast, the addition of silane-modified sepiolite to the composite resulted in better dispersion, a higher content of bound rubber, a stronger interfacial bonding ability, and a higher crosslinking density, resulting in a lower stress decrease and a higher equilibrium stress value. The trend of stress reduction for the composites modified with silane-modified sepiolites follows VSep/NR > Msep/NR > Tsep/NR, which is the opposite to the trend of bound rubber content and crosslinking density. This indicates the difference in network structures created by different silane-modified sepiolites within the composite.

Table 5. Static mechanical properties of sepiolite/NR composites.

Samples	Modulus at 100%/MPa	Modulus at 300%/MPa	Tensile Strength/MPa	Elongation at Break/%	Tear Strength/N·mm^{-1}
Sep/NR	3.45 ± 0.33	8.82 ± 0.88	30.0 ± 1.8	618 ± 58	49.6 ± 1.7
VSep/NR	3.61 ± 0.27	10.28 ± 0.91	29.98 ± 2.1	640 ± 47	50.3 ± 2.0
MSep/NR	4.14 ± 0.15	14.99 ± 0.53	23.4 ± 1.1	476 ± 35	58.1 ± 1.5
TSep/NR	4.43 ± 0.19	16.87 ± 0.62	25.5 ± 1.6	450 ± 36	60.3 ± 1.4

As shown in Figure 10, the sepiolite grafted with VTES primarily consists of the C=C functional group, which relies on the addition of sulfur to form a chemical bond with the rubber molecular chain, as well as better dispersion, which increases the bound rubber content and crosslinking density. The sepiolite grafted with MPTES and TESPT not only has good dispersion, a large grafting molecular weight, and a more entangled rubber molecular chain, but can also form chemical bonds with rubber molecular chains via its own -HS and $-S_4-$ to form a denser bonded rubber and network crosslinking structure. Table 3 shows that TESPT-Sep grafting had the largest amount and the most -S- content, resulting in a more denser crosslinking network structure and stronger physical properties in TSep/NR composites.

3.5. Dynamic Properties of Sepiolite/NR Composites

The dynamic mechanical properties of a composite reflect the amount of energy stored as elastic energy and the amount of energy dissipated during strain, which are highly dependent on the volume fraction of the filler, its dispersion in the matrix, and the interfacial bonding between the filler and the matrix [64]. Figure 11A shows the temperature-dependent storage modulus (E') of sepiolite/NR composites, which indicates that the addition of silane-modified sepiolite leads to higher E'. At 25 °C, the E' values of TSep/NR and MSep/NR were 8.55 MPa and 8.18 MPa, respectively, representing an improvement of 45.4% and 39.1% over Sep/NR. This result suggests that silane modification induces stronger interfacial interactions between sepiolite and natural rubber matrix. Figure 11B shows the temperature dependence of loss factor (tan δ) of the sepiolite/NR composites, and all samples exhibited an obvious loss peak at around −40 °C, which corresponds to the glass transition temperature (T_g) of the composites. The addition of silane-modified sepiolite led to a clear shift to higher T_g values, indicating that more polymer chains were grafted or adsorbed on the fillers, slowing down polymer kinetics and increasing the Tg of the composite. Additionally, the peak of tan δ tended to decrease due to the stronger interfacial interaction between the modified sepiolite and the rubber matrix and the more dense crosslinked network structure, which produced a larger E' (tan δ = E''/E').

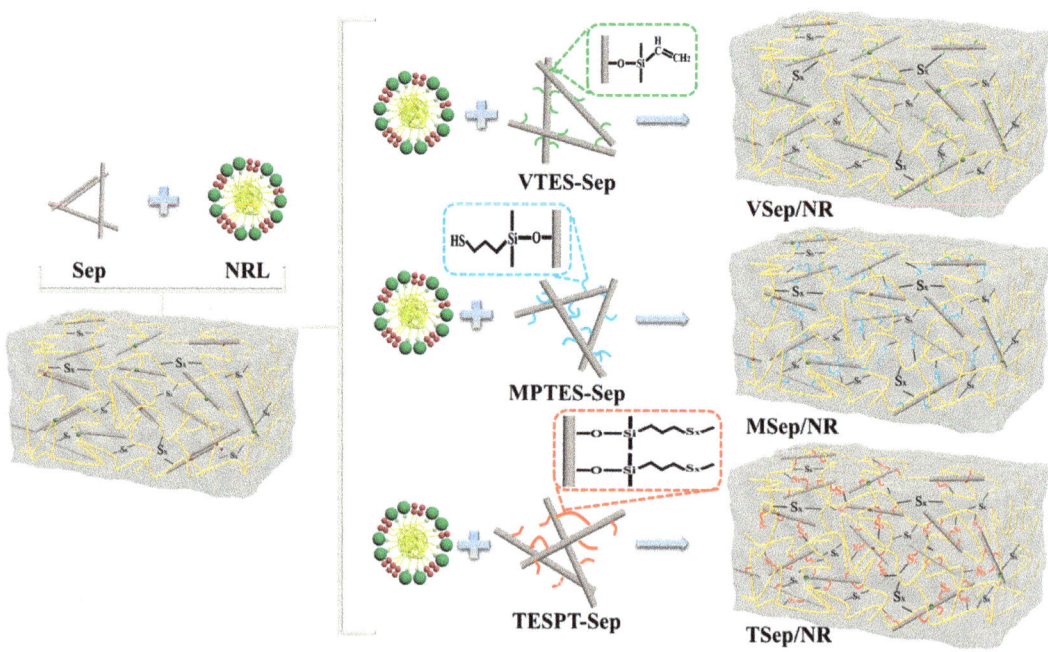

Figure 10. Schematic diagram of the structure of different modified sepiolite-reinforced natural rubbers.

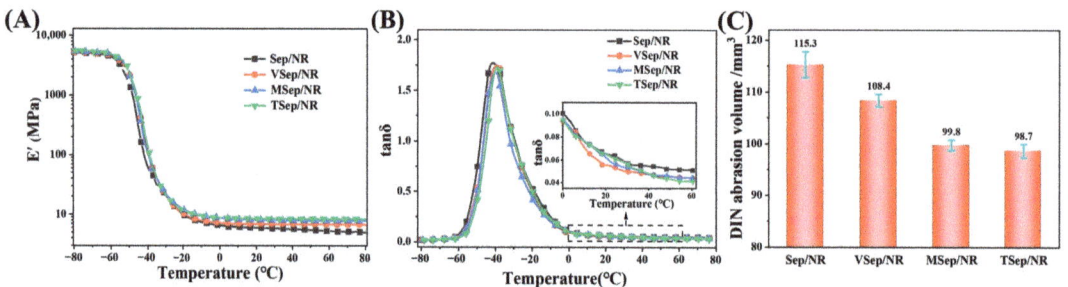

Figure 11. (**A**) Temperature dependence of the E', (**B**) tan δ, and (**C**) DIN abrasion volume of sepiolite/NR composites.

In tire tread rubber, the tan δ at 0 °C and 60 °C are important parameters related to the wet skid resistance and the rolling resistance. High-performance rubber composites should have high tan δ 0 °C and low tan δ at 60 °C [65]. From Figure 11B, it is evident that the tan δ of composites with silane-modified sepiolite at both 0 °C and 60 °C were smaller than those of Sep/NR, indicating reduced wet skid resistance and improved rolling resistance. In addition, Figure 11C indicates that the DIN abrasion volume of composites with silane-modified sepiolite was reduced compared to Sep/NR, indicating improved abrasion resistance. In summary, the use of silane-modified sepiolite can improve the interfacial interaction between sepiolite and the rubber matrix, enhance the abrasion resistance, and improve the rolling resistance of the composite, but the wet skid resistance may be reduced.

4. Conclusions

This study presented an innovative in situ modification method for sepiolite to be used in latex compounding. Three types of silane coupling agents were successfully grafted onto the sepiolite surface during the disaggregation and activation of natural sepiolite, which was confirmed by FTIR, TG, and SEM analyses. This fabrication technique is both efficient and easy to operate. The silane coupling agents grafted onto the sepiolite surface promoted the dispersion of sepiolite in the natural rubber matrix and enhanced the interfacial bonding strength between sepiolite and natural rubber by increasing the entangled rubber molecular chains and chemical bonding interactions. VTES-modified sepiolite, which contains C=C, retarded the vulcanization and slightly improved the physical and mechanical properties of the composites. MPTES-modified sepiolite and TESPT-modified sepiolite, which both contain -S-, effectively accelerated vulcanization and led to a denser crosslinked network structure, resulting in stronger physical and mechanical properties of the composites. TESPT had the highest grafting amount, and the modulus at 300% increased from 8.82 MPa to 16.87 MPa, tear strength increased from 49.6 $N\cdot mm^{-1}$ to 60.3 $N\cdot mm^{-1}$, and the rolling resistance and abrasive resistance of the composites improved.

Author Contributions: Z.H., conceptualization, methodology, software, formal analysis, investigation, data curation, writing—original draft preparation, supervision; D.Z., validation, formal analysis, investigation, data curation, writing—original draft preparation; Q.C., conceptualization, methodology, formal analysis, investigation, resources, writing—review and editing, visualization, supervision, project administration, funding acquisition; Z.X., methodology, project administration, funding acquisition. All authors have read and agreed to the published version of the manuscript.

Funding: This research was funded by the National Key R&D Program of China under grant number 2022YFD2301202.

Institutional Review Board Statement: Not applicable.

Data Availability Statement: The data that support the findings of this study are available from the corresponding author upon reasonable request.

Conflicts of Interest: The authors declare no conflict of interest.

References

1. Dunuwila, P.; Rodrigo, V.H.L.; Goto, N. Sustainability of natural rubber processing can be improved: A case study with crepe rubber manufacturing in Sri Lanka. *Resour. Conserv. Recycl.* **2018**, *133*, 417–427. [CrossRef]
2. Dominic, M.; Joseph, R.; Sabura Begum, P.M.; Kanoth, B.P.; Chandra, J.; Thomas, S. Green tire technology: Effect of rice husk derived nanocellulose (RHNC) in replacing carbon black (CB) in natural rubber (NR) compounding. *Carbohydr. Polym.* **2020**, *230*, 115620. [CrossRef]
3. Dwivedi, C.; Manjare, S.; Rajan, S.K. Recycling of waste tire by pyrolysis to recover carbon black: Alternative & environment-friendly reinforcing filler for natural rubber compounds. *Compos. Part B Eng.* **2020**, *200*, 108346. [CrossRef]
4. Fan, Y.; Fowler, G.D.; Zhao, M. The past, present and future of carbon black as a rubber reinforcing filler—A review. *J. Clean. Prod.* **2020**, *247*, 119115. [CrossRef]
5. Kato, A.; Ikeda, Y.; Kohjiya, S. Reinforcement Mechanism of Carbon Black (CB) in Natural Rubber Vulcanizates: Relationship Between CB Aggregate and Network Structure and Viscoelastic Properties. *Polym.-Plast. Technol. Eng.* **2018**, *57*, 1418–1429. [CrossRef]
6. Gui, Y.; Zheng, J.; Ye, X.; Han, D.; Xi, M.; Zhang, L. Preparation and performance of silica/SBR masterbatches with high silica loading by latex compounding method. *Compos. Part B Eng.* **2016**, *85*, 130–139. [CrossRef]
7. Ryu, C.S.; Kim, K.-J. Interfacial Adhesion in Silica-Silane Filled NR Composites: A Short Review. *Polymers* **2022**, *14*, 2705. [CrossRef] [PubMed]
8. Errington, E.; Guo, M.; Heng, J.Y.Y. Synthetic amorphous silica: Environmental impacts of current industry and the benefit of biomass-derived silica. *Green Chem.* **2023**. [CrossRef]
9. Matos, C.F.; Galembeck, F.; Zarbin, A.J.G. Multifunctional and environmentally friendly nanocomposites between natural rubber and graphene or graphene oxide. *Carbon* **2014**, *78*, 469–479. [CrossRef]
10. Alimardani, M.; Abbassi-Sourki, F.; Bakhshandeh, G.R. An investigation on the dispersibility of carbon nanotube in the latex nanocomposites using rheological properties. *Compos. Part B Eng.* **2014**, *56*, 149–156. [CrossRef]

11. Xu, Z.; Jerrams, S.; Guo, H.; Zhou, Y.; Jiang, L.; Gao, Y.; Zhang, L.; Liu, L.; Wen, S. Influence of graphene oxide and carbon nanotubes on the fatigue properties of silica/styrene-butadiene rubber composites under uniaxial and multiaxial cyclic loading. *Int. J. Fatigue* **2020**, *131*, 105388. [CrossRef]
12. Cao, L.; Huang, J.; Chen, Y. Dual Cross-linked Epoxidized Natural Rubber Reinforced by Tunicate Cellulose Nanocrystals with Improved Strength and Extensibility. *ACS Sustain. Chem. Eng.* **2018**, *6*, 14802–14811. [CrossRef]
13. Hakimi, N.M.F.; Lee, S.H.; Lum, W.C.; Mohamad, S.F.; Osman Al Edrus, S.S.; Park, B.-D.; Azmi, A. Surface Modified Nanocellulose and Its Reinforcement in Natural Rubber Matrix Nanocomposites: A Review. *Polymers* **2021**, *13*, 3241. [CrossRef] [PubMed]
14. Alwis, G.M.C.; Kottegoda, N.; Ratnayake, U.N. Facile exfoliation method for improving interfacial compatibility in montmorillonite-natural rubber nanocomposites: A novel charge inversion approach. *Appl. Clay Sci.* **2020**, *191*, 105633. [CrossRef]
15. Zha, C.; Wang, W.; Lu, Y.; Zhang, L. Constructing Covalent Interface in Rubber/Clay Nanocomposite by Combining Structural Modification and Interlamellar Silylation of Montmorillonite. *ACS Appl. Mater. Interfaces* **2014**, *6*, 18769–18779. [CrossRef] [PubMed]
16. Yang, Y.; Zhang, H.; Zhang, K.; Liu, L.; Ji, L.; Liu, Q. Vulcanization, interfacial interaction, and dynamic mechanical properties of in-situ organic amino modified kaolinite/SBR nanocomposites based on latex compounding method. *Appl. Clay Sci.* **2020**, *185*, 105366. [CrossRef]
17. Komadel, P.; Madejová, J.; Stucki, J.W. Structural Fe(III) reduction in smectites. *Appl. Clay Sci.* **2006**, *34*, 88–94. [CrossRef]
18. Guggenheim, S.; Krekeler, M.P.S. Chapter 1—The Structures and Microtextures of the Palygorskite–Sepiolite Group Minerals. In *Developments in Clay Science*; Galàn, E., Singer, A., Eds.; Elsevier: Amsterdam, The Netherlands, 2011; Volume 3, pp. 3–32.
19. Suárez, M.; García-Romero, E. Chapter 2—Advances in the Crystal Chemistry of Sepiolite and Palygorskite. In *Developments in Clay Science*; Galàn, E., Singer, A., Eds.; Elsevier: Amsterdam, The Netherlands, 2011; Volume 3, pp. 33–65.
20. Nagy, B.; Bradley, W.F. The Structural Scheme of Sepiolite. *Clay Miner. Bull.* **1954**, *2*, 203. [CrossRef]
21. Tian, G.; Han, G.; Wang, F.; Liang, J. 3—Sepiolite Nanomaterials: Structure, Properties and Functional Applications. In *Nanomaterials from Clay Minerals*; Wang, A., Wang, W., Eds.; Elsevier: Amsterdam, The Netherlands, 2019; pp. 135–201.
22. Erdoğan Alver, B. Hydrogen adsorption on natural and sulphuric acid treated sepiolite and bentonite. *Int. J. Hydrog. Energy* **2018**, *43*, 831–838. [CrossRef]
23. Zhou, F.; Yan, C.; Zhang, Y.; Tan, J.; Wang, H.; Zhou, S.; Pu, S. Purification and defibering of a Chinese sepiolite. *Appl. Clay Sci.* **2016**, *124–125*, 119–126. [CrossRef]
24. Zhuang, G.; Gao, J.; Chen, H.; Zhang, Z. A new one-step method for physical purification and organic modification of sepiolite. *Appl. Clay Sci.* **2018**, *153*, 1–8. [CrossRef]
25. Leblanc, J.L. Rubber–filler interactions and rheological properties in filled compounds. *Prog. Polym. Sci.* **2002**, *27*, 627–687. [CrossRef]
26. Di Credico, B.; Tagliaro, I.; Cobani, E.; Conzatti, L.; D'Arienzo, M.; Giannini, L.; Mascotto, S.; Scotti, R.; Stagnaro, P.; Tadiello, L. A Green Approach for Preparing High-Loaded Sepiolite/Polymer Biocomposites. *Nanomaterials* **2019**, *9*, 46. [CrossRef] [PubMed]
27. Carignani, E.; Cobani, E.; Martini, F.; Nardelli, F.; Borsacchi, S.; Calucci, L.; Di Credico, B.; Tadiello, L.; Giannini, L.; Geppi, M. Effect of sepiolite treatments on the oxidation of sepiolite/natural rubber nanocomposites prepared by latex compounding technique. *Appl. Clay Sci.* **2020**, *189*, 105528. [CrossRef]
28. Hou, C.; Gao, L.; Yu, H.; Sun, Y.; Yao, J.; Zhao, G.; Liu, Y. Preparation of magnetic rubber with high mechanical properties by latex compounding method. *J. Magn. Magn. Mater.* **2016**, *407*, 252–261. [CrossRef]
29. Fernandes, F.M.; Ruiz-Hitzky, E. Assembling nanotubes and nanofibres: Cooperativeness in sepiolite–carbon nanotube materials. *Carbon* **2014**, *72*, 296–303. [CrossRef]
30. Ruiz-Hitzky, E.; Ruiz-García, C.; Fernandes, F.M.; Lo Dico, G.; Lisuzzo, L.; Prevot, V.; Darder, M.; Aranda, P. Sepiolite-Hydrogels: Synthesis by Ultrasound Irradiation and Their Use for the Preparation of Functional Clay-Based Nanoarchitected Materials. *Front. Chem.* **2021**, *9*, 733105. [CrossRef]
31. Jiménez-López, A.; López-González, J.d.D.; Ramīrez-Sāenz, A.; Rodrīguez-Reinoso, F.; Valenzuela-Calahorro, C.; Zurita-Herrera, L. Evolution of surface area in a sepiolite as a function of acid and heat treatments. *Clay Miner.* **1978**, *13*, 375–385. [CrossRef]
32. Hayeemasae, N.; Ismail, H. Reinforcement of epoxidized natural rubber through the addition of sepiolite. *Polym. Compos.* **2019**, *40*, 924–931. [CrossRef]
33. Raji, M.; Mekhzoum, M.E.M.; Rodrigue, D.; Qaiss, A.e.k.; Bouhfid, R. Effect of silane functionalization on properties of polypropylene/clay nanocomposites. *Compos. Part B Eng.* **2018**, *146*, 106–115. [CrossRef]
34. Peinado, V.; García, L.; Fernández, Á.; Castell, P. Novel lightweight foamed poly(lactic acid) reinforced with different loadings of functionalised Sepiolite. *Compos. Sci. Technol.* **2014**, *101*, 17–23. [CrossRef]
35. Wang, F.; Feng, L.; Tang, Q.; Liang, J.; Liu, H.; Liu, H. Effect of Modified Sepiolite Nanofibers on Properties of cis-Polybutadiene Rubber Composite Nanomaterials. *J. Nanomater.* **2013**, *2013*, 369409.
36. Mohd Zaini, N.A.; Samsudin, D.; Rusli, A.; Ismail, H. Curing, thermal, tensile and flammability characteristics of sepiolite/ethylene propylene diene monomer rubber composites with glut palmitate salt and silane coupling agents. *Polym. Compos.* **2022**, *43*, 4721–4736. [CrossRef]

37. López Valentín, J.; Rodríguez Díaz, A.; Ibarra Rueda, L.; González Hernández, L. Effect of a natural magnesium silicate treated with a new coupling agent on the properties of ethylene–propylene–diene rubber compounds. *J. Appl. Polym. Sci.* **2004**, *91*, 1489–1493. [CrossRef]
38. Kim, K.-J.; Vanderkooi, J. Moisture effects on improved hydrolysis reaction for TESPT and TESPD-silica compounds. *Compos. Interfaces* **2004**, *11*, 471–488. [CrossRef]
39. Yin, C.; Zhang, Q.; Liu, J.; Liu, L.; Gu, J. Preparation, properties of In-situ silica modified styrene-butadiene rubber and its silica-filled composites. *Polym. Compos.* **2018**, *39*, 22–28. [CrossRef]
40. Hayeemasae, N.; Masa, A.; Othman, N.; Surya, I. Viable Properties of Natural Rubber/Halloysite Nanotubes Composites Affected by Various Silanes. *Polymers* **2023**, *15*, 29. [CrossRef]
41. Li, Q.; Li, X.; Lee, D.-H.; Fan, Y.; Nam, B.-U.; Lee, J.-E.; Cho, U.-R. Hybrid of bamboo charcoal and silica by tetraethoxysilane hydrolysis over acid catalyst reinforced styrene-butadiene rubber. *J. Appl. Polym. Sci.* **2018**, *135*, 46219. [CrossRef]
42. Vollet, D.R.; Barreiro, L.A.; Paccola, C.E.T.; Awano, C.M.; De Vicente, F.S.; Yoshida, M.; Donatti, D.A. A kinetic modeling for the ultrasound-assisted and oxalic acid-catalyzed hydrolysis of 3-glycidoxypropyltrimethoxysilane. *J. Sol-Gel Sci. Technol.* **2016**, *80*, 873–880. [CrossRef]
43. Negrete; Letoffe, J.-M.; Putaux, J.-L.; David, L.; Bourgeat-Lami, E. Aqueous Dispersions of Silane-Functionalized Laponite Clay Platelets. A First Step toward the Elaboration of Water-Based Polymer/Clay Nanocomposites. *Langmuir* **2004**, *20*, 1564–1571. [CrossRef]
44. Zhong, B.; Jia, Z.; Luo, Y.; Guo, B.; Jia, D. Preparation of halloysite nanotubes supported 2-mercaptobenzimidazole and its application in natural rubber. *Compos. Part A Appl. Sci. Manuf.* **2015**, *73*, 63–71. [CrossRef]
45. Sun, Y.; Cheng, Z.; Zhang, L.; Jiang, H.; Li, C. Promoting the dispersibility of silica and interfacial strength of rubber/silica composites prepared by latex compounding. *J. Appl. Polym. Sci.* **2020**, *137*, 49526. [CrossRef]
46. Flory, P.J. Statistical Mechanics of Swelling of Network Structures. *J. Chem. Phys.* **1950**, *18*, 108–111. [CrossRef]
47. *ISO 37:2005*; Rubber, Vulcanized or Thermoplastic—Determination of Tensile Stress-Strain Properties. ISO: Geneva, Switzerland, 2005.
48. *ISO 34-1:2015*; Rubber, Vulcanized or Thermoplastic—Determination of Tear Strength—Part 1: Trouser, Angle and Crescent Test Pieces. ISO: Geneva, Switzerland, 2015.
49. *ISO 4649:2017*; Rubber, Vulcanized or Thermoplastic—Determination of Abrasion Resistance Using a Rotating Cylindrical Drum Device. ISO: Geneva, Switzerland, 2017.
50. Tartaglione, G.; Tabuani, D.; Camino, G. Thermal and morphological characterisation of organically modified sepiolite. *Microporous Mesoporous Mater.* **2008**, *107*, 161–168. [CrossRef]
51. Doğan, M.; Turhan, Y.; Alkan, M.; Namli, H.; Turan, P.; Demirbaş, Ö. Functionalized sepiolite for heavy metal ions adsorption. *Desalination* **2008**, *230*, 248–268. [CrossRef]
52. Yan, W.; Liu, D.; Tan, D.; Yuan, P.; Chen, M. FTIR spectroscopy study of the structure changes of palygorskite under heating. *Spectrochim. Acta Part A Mol. Biomol. Spectrosc.* **2012**, *97*, 1052–1057. [CrossRef] [PubMed]
53. Abali, S.; Aydin, Y.A. Silanization of sepiolite with various silane coupling agents for enhancing oil uptake capacity. *Sep. Sci. Technol.* **2023**, *58*, 586–597. [CrossRef]
54. Duquesne, E.; Moins, S.; Alexandre, M.; Dubois, P. How can Nanohybrids Enhance Polyester/Sepiolite Nanocomposite Properties? *Macromol. Chem. Phys.* **2007**, *208*, 2542–2550. [CrossRef]
55. Ruiz, R.; del Moral, J.C.; Pesquera, C.; Benito, I.; González, F. Reversible folding in sepiolite: Study by thermal and textural analysis. *Thermochim. Acta* **1996**, *279*, 103–110. [CrossRef]
56. Nagata, H.; Shimoda, S.; Sudo, T. On Dehydration of Bound Water of Sepiolite. *Clays Clay Miner.* **1974**, *22*, 285–291. [CrossRef]
57. García, N.; Guzmán, J.; Benito, E.; Esteban-Cubillo, A.; Aguilar, E.; Santarén, J.; Tiemblo, P. Surface Modification of Sepiolite in Aqueous Gels by Using Methoxysilanes and Its Impact on the Nanofiber Dispersion Ability. *Langmuir* **2011**, *27*, 3952–3959. [CrossRef] [PubMed]
58. Payne, A.R.; Whittaker, R.E. Low Strain Dynamic Properties of Filled Rubbers. *Rubber Chem. Technol.* **1971**, *44*, 440–478. [CrossRef]
59. Wang, Z.-P.; Zhang, H.; Liu, Q.; Wang, S.-J.; Yan, S.-K. Effects of Interface on the Dynamic Hysteresis Loss and Static Mechanical Properties of Illite Filled SBR Composites. *Chin. J. Polym. Sci.* **2022**, *40*, 1493–1502. [CrossRef]
60. Yan, H.; Sun, K.; Zhang, Y.; Zhang, Y.; Fan, Y. Effects of silane coupling agents on the vulcanization characteristics of natural rubber. *J. Appl. Polym. Sci.* **2004**, *94*, 1511–1518. [CrossRef]
61. Rooj, S.; Das, A.; Thakur, V.; Mahaling, R.N.; Bhowmick, A.K.; Heinrich, G. Preparation and properties of natural nanocomposites based on natural rubber and naturally occurring halloysite nanotubes. *Mater. Des.* **2010**, *31*, 2151–2156. [CrossRef]
62. He, S.; Xue, Y.; Lin, J.; Zhang, L.; Du, X.; Chen, L. Effect of silane coupling agent on the structure and mechanical properties of nano-dispersed clay filled styrene butadiene rubber. *Polym. Compos.* **2016**, *37*, 890–896. [CrossRef]
63. Lipińska, M.; Imiela, M. Morphology, rheology and curing of (ethylene-propylene elastomer/hydrogenate acrylonitrile-butadiene rubber) blends reinforced by POSS and organoclay. *Polym. Test.* **2019**, *75*, 26–37. [CrossRef]

64. Wu, X. Natural rubber/graphene oxide composites: Effect of sheet size on mechanical properties and strain-induced crystallization behavior. *Express Polym. Lett.* **2015**, *9*, 672–685. [CrossRef]
65. Cao, L.; Sinha, T.K.; Tao, L.; Li, H.; Zong, C.; Kim, J.K. Synergistic reinforcement of silanized silica-graphene oxide hybrid in natural rubber for tire-tread fabrication: A latex based facile approach. *Compos. Part B Eng.* **2019**, *161*, 667–676. [CrossRef]

Disclaimer/Publisher's Note: The statements, opinions and data contained in all publications are solely those of the individual author(s) and contributor(s) and not of MDPI and/or the editor(s). MDPI and/or the editor(s) disclaim responsibility for any injury to people or property resulting from any ideas, methods, instructions or products referred to in the content.

Article

ESBR Nanocomposites Filled with Monodisperse Silica Modified with Si747: The Effects of Amount and pH on Performance

Lijian Xia [1,2], Anmin Tao [1], Jinyun Cui [1], Abin Sun [1], Ze Kan [1,*] and Shaofeng Liu [1]

[1] Key Laboratory of Biobased Polymer Materials, Shandong Provincial Education Department, School of Polymer Science and Engineering, Qingdao University of Science and Technology, Qingdao 266042, China

[2] State Key Laboratory of Marine Coatings, Marine Chemical Research Institute Co., Ltd., Qingdao 266072, China

* Correspondence: zkan@qust.edu.cn

Abstract: To prepare silica/rubber composites for low roll resistance tires, a novel strategy was proposed in this study, in which autonomous monodisperse silica (AS) was prepared and modified using 3-mercaptopropyloxy-methoxyl-bis(nonane-pentaethoxy) siloxane (Si747), after which silica/emulsion styrene butadiene rubber (ESBR) master batches were produced using the latex compounding technique. Meanwhile, the commercial precipitated silica (PS) was introduced as a control. In this study, the effects of amount of Si747 and pH value on the properties of the silica/ESBR composites were systematically analyzed. Thermal gravimetric analysis (TGA) and Fourier transform infrared (FTIR) results indicated that Si747 reduced the silanol group by chemical grafting and physical shielding, and the optimum amounts of Si747 for AS and PS modification were confirmed to be 15% and 20%, respectively. Under a pH of 9, ESBR/modified AS (MAS) composites with 15% Si747 presented better silica dispersion and a weaker Payne effect, compared with ESBR/modified PS (MPS) composites with 20% Si747. Meanwhile, in terms of dynamic properties, the ESBR/MAS composites exhibited a better balance of lower rolling resistance and higher wet skid resistance than the ESBR/MPS composites.

Keywords: monodisperse silica; latex compounding technique; Si747; ESBR/silica nanocomposites

Citation: Xia, L.; Tao, A.; Cui, J.; Sun, A.; Kan, Z.; Liu, S. ESBR Nanocomposites Filled with Monodisperse Silica Modified with Si747: The Effects of Amount and pH on Performance. *Polymers* 2023, 15, 981. https://doi.org/10.3390/polym15040981

Academic Editor: Andrea Sorrentino

Received: 25 December 2022
Revised: 7 February 2023
Accepted: 10 February 2023
Published: 16 February 2023

Copyright: © 2023 by the authors. Licensee MDPI, Basel, Switzerland. This article is an open access article distributed under the terms and conditions of the Creative Commons Attribution (CC BY) license (https://creativecommons.org/licenses/by/4.0/).

1. Introduction

Silica is an extremely important reinforcing filler in the rubber industry [1–3]. Therein, silica/styrene butadiene rubber (SBR) composites are usually used for green tires because silica provides a much better combination of low rolling resistance and considerably high wet skid resistance than composites of carbon black fills [3–5]. However, silica features high polarity and strong hydrophilicity because numerous hydroxyl groups exist on the silica surface, resulting in serious aggregation and poor compatibility with non-polar rubber and poor dispersion in the rubber matrix [6–8]. In order to improve the dispersion of silica, one major method is to consume or shield the hydroxyl groups and form a "coupling bridge" between the silica and the rubber by introducing different kinds of silane coupling agent [9–11]. Among them, bis(3-triethoxy-silylpropyl) tetrasulfide (TESPT, Si69) is the most commonly used. Li used a "two-step method" to investigate the modification process and elaborated the modification mechanism in detail based on the traditional method of blending rubber, silica and Si69 to prepare compound [12].

The latex compounding technique is another common method developed for improving silica dispersion and lowering the processing temperature in traditional mechanical blending. Many attempts have been made to explore feasible and advantageous technologies using the latex compounding technique, including preparing in situ and then co-flocculating silica particles using the sol–gel method in rubber latex [13–15], or mechanical stirring of commercial silica slurry followed by blending with rubber latex [16].

In our previous report [17], we prepared autonomous monodisperse silica (AS) via the sol–gel method and precisely controlled the morphology and particle size, then prepared Natural Rubber (NR)/AS master batches via the latex compounding technique. Meanwhile, commercial precipitated silica (PS) was introduced as a control. NR/AS composites exhibited better silica dispersion and weaker filler–filler interactions compared to NR/PS composites. However, it should be noted that the process of modifying AS and PS with Si69 was inefficient and time-consuming, and Si69, due to its abundant polysulfide bonds, can more easily cause scorching of rubber during the compounding process along with poor storage safety. Moreover, the amounts of ethanol aggravated volatile organic compounds (VOC) emission. In summary, the research and development of silane coupling agents with low sulfur (*Bis*(triethoxysilylpropyl)disulphide, TESPD, Si75, Si266) [18,19] or shielding sulfur (Octanoyl thioester-protected mercaptosilane, NXT) [20] are of great interest. Both Si75 and NXT decrease reactivity and ensure process safety; NXT also reduces VOC production and improves the rolling resistance of compounds. Due to the strict requirements of EU labeling law for VOC and the higher price of NXT, a kind of silane coupling agent where the long-chain polyalkyl ether alcohols on the silane's silicon atom replace the ethoxyl is used instead, resulting in greatly reduced VOC emission. In this case, long-chain polyalkyl ether alcohols can shield the reactive sulfhydryl group and delay the activation of the accelerant and sulfur. In addition, long-chain polyalkyl ether alcohols with high polarity can adsorb on the silica surface, thus weakenig the silica agglomerate and the filler–filler network. Silane coupling agents of this type include 3-mercaptopropyloxy-ethyoxyl-*bis*(tridecyl-pentaethoxy) siloxane (Si363) [21] and 3-mercaptopropyloxy-methyoxyl-*bis*(nonane-pentaethoxy) siloxane (Si747) [22]. Si363 and Si747, as water-soluble silane coupling agents, hydrolyze easily in water and modify silica directly in its solution state. The modified silica/rubber composite, with high scorch resistance, has higher processing safety as a result. Si747 has a significant cost advantage compared to the imported product Si363.

In this article, we explored the mechanism of interaction between Si747 and monodisperse or precipitated silica. Other researchers reported that the amount of silane coupling agent [12] and modification conditions [23,24] (temperature, pH) were important for the degree of silica surface modification. In particular, the hydrolysis and condensation reaction of the silane coupling agent is affected by the structure of the hydrolysis group, the reaction medium and the reaction conditions (temperature, pH, concentration, amount of water and catalyst) [25]. Pantoja [26] proved that pH has a great influence on the hydrolysis reaction of silane coupling agent γ-methacryloxypropyltrimethoxysilane (MPS) via infrared spectroscopy. Rostami [27] studied the effect of different pH on the surface chemistry of fumed silica modified by aminopropyltrimethoxysilane (APTMS). In this research study, we confirmed the optimum amount of Si747 and modified pH in terms of modification efficiency of AS and PS, respectively, using Fourier transform infrared (FT-IR) and thermal weight loss analysis (TGA), after which silica/ESBR master batches were prepared via the latex compounding technique. Finally, the properties of the silica/ESBR composites were investigated.

2. Materials and Methods

2.1. Materials

L-lysine (98%) was purchased from Aladdin Industrial Corporation (Shanghai, China); Ttraethoxysilane (TEOS, 98%) was produced by Sinopharm Chemical Reagent Co., Ltd. (Shanghai, China); Si747 was produced by Shanghai Cheeshine Chemicals Co., Ltd. (Shanghai, China); Commercial ESBR1502 latex with 23.32% mass fraction dry rubber content was produced by Sinopec Qilu Petrochemical Co., Ltd. (Zibo, China). Commercial PS (1165 MP, BET surface area 165 m^2/g) was purchased from Solvay white carbon black of Qingdao Co., Ltd. (Qingdao, China). All of the rubber ingredients were industrial grade and used as received.

2.2. Preparation of Modified Silica

Specific amounts of AS (50 g) with desired sizes (28 nm, BET surface area 169 m^2/g) were obtained via the method reported by Yokoi [28], using L-lysine as the catalyst. The synthetic silica suspension was concentrated to a volume of 150 mL using a rotary evaporator and reserved. As control, 50 g carefully weighed PS was added into the same volume (150 mL) of deionized water and stirred at 300 rpm for 30 min before use.

In our typical experiments, hydrochloric acid, saturated sodium bicarbonate solution and saturated sodium hydroxide were used to adjust the pH values to 3, 7, 9, and 12, respectively. Different masses of Si747 (mass ratio of Si747 to water = 11.11% and mass ratios of Si747 to silica = 8%, 10%, 12%, 15%, 20%, respectively) were added into deionized water for hydrolysis for 12 h at room temperature.

The pre-prepared AS suspension and PS slurry were then mixed with different amounts of Si747 hydrolysates. The volumes of all these mixed solutions were no more than 250 mL, and were stirred at 600 rpm and 80 °C for 6 h in an oil bath. For convenience of description, the modified silica solutions are denoted 8%-AS, 10%-AS, 12%-AS, 15%-AS, 20%-AS, 10%-PS, 12%-PS, 15%-PS, 20%-PS, 15%-AS@3, 15%-AS@7, 15%-AS@9, 15%-AS@12, 20%-PS@7, and 20%-PS@9.

Part of the pure silica and modified silica suspension were placed into a drying oven under 110 °C for 12 h. Silica powders were extracted in a Soxhlet extractor using ethanol for 24 h (110 °C, 30 min for each reflux) to remove the self-condensed Si747. Then, all silica powders after extraction were put into a vacuum drying oven at 65 °C for 24 h.

2.3. Preparation of Silica/ESBR Master Batches

Different types (monodisperse, precipitated, modification at different conditions) of silica solutions were agitated for 10 min using a mechanical stirrer and blended with the weighed ESBR latex. After stirring together for 30 min at 300 rpm with a mechanical stirrer, the master batches of silica/ESBR were co-flocculated using calcium ethylate (3 g calcium nitrate in 97 g ethyl alcohol). The flocculates were made to sheets manually and the above steps were repeated until the latex was completely demulsified. All master batch sheets were washed with water for several times. The residual solution was centrifuged and the solids and master batch sheets were then collected to dry at 55 °C in the oven until a constant weight was reached.

2.4. Preparation of Silica/ESBR Compounds and Vulcanizates

In order to obtain well-dispersed silica/ESBR compounds, two stages of mixing were carried out. First, silica/ESBR master batches were masticated for 3 min in a torque rheometer (RM-200C, Harbin Harper Electric Technology Co., Ltd., Harbin, China) for initial mixing, for which the initial temperature was set at 90 °C and the rotational speed was constant at 600 rpm. When the torque curve was stable, the compression lever was lifted. Following the formulation listed in Table 1, zinc oxide and stearic acid were added at the same time. After mixing for 8 min, the torque curve was stable and the temperature of the chamber reached 130 °C; the compounds were then taken out.

Next, N-tert-butylbenzothiazole-2-sulfonamide (NS), diphenyl guanidine (DPG) and sulfur (S) were successively added into the cooled compounds on a 6-inch two-roll mill (Dongguan Bolon Precision Testing Machines Co., Ltd., Dongguan, China), blending uniformly at room temperature. Condensate water was flowed into the roller constantly to maintain a temperature below 55 °C. The whole mixing process took 15 min for each sample and compound sheets with an approximate thickness of 2 mm were obtained.

The silica/ESBR compounds are denoted 8%-AS-R, 10%-AS-R, 12%-AS-R, 15%-AS-R 20%-AS-R, 10%-PS-R, 12%-PS-R, 15%-PS-R, 20%-PS-R, 15%-AS@3-R, 15%-AS@7-R, 15%-AS@9-R, 15%-AS@12-R, 20%-PS@7-R, and 20%-PS@9-R, based on the silica type and modification conditions.

Table 1. Formulation of Silica/ESBR Compounds.

Material	Content/phr [1]
Dried ESBR [2]	100
AS	50
PS	50
Si747	Variable [3]
Zinc oxide [4]	3
Stearic acid [5]	1
N-tert-butylbenzothiazole-2-sulfonamide [6]	1.5
Diphenyl guanidine [7]	1.5
Sulfur [8]	1.75

[1.] Parts per hundred of rubber. [2.] ESBR latex was coagulated and dried, then weighed the mass. [3.] The amount of Si747 was calculated based on the mass of silica and the modification conditions could be variable. [4.] Zinc oxide was used to activate the whole vulcanization system and improve the crosslinking density and aging resistance of vulcanized rubber. [5.] Stearic acid can be used as plasticizer and is conducive to the full dispersion of silica, while reacting with zinc oxide to promote vulcanization. [6.] N-tert-butylbenzothiazole-2-sulfonamide was used as the after-effect vulcanization accelerator. [7.] Diphenyl guanidine was used as the medium speed vulcanization accelerator. [8.] Sulfur was used as the cross-linking agent.

The scorch time (t_{s2}) and optimum cure time (t_{90}) of the silica/ESBR compounds were measured using a rheometer vulcanization machine (MDR-2000, Alpha, Akron, OH, USA) at 160 °C after being stored at room temperature for 12 h. The oscillating frequency was 1.7 ± 0.1 Hz, with an amplitude of ±3°. The volume of each test specimen was 5 cm^3. The cure rate index (CRI) was calculated using the following equation:

$$\text{CRI (min}^{-1}\text{)} = 100/(t_{90} - t_{s2}), \tag{1}$$

The compounds were vulcanized at 160 °C and 10 MPa for (t_{90} + 2) min in a hydraulic press (HS100T-RTMO, Shenzhen Jiaxin Co., Ltd., Shenzhen, China).

2.5. Characterization

SEM photographs of the silica suspensions were taken on a Quanta FEG250 field emission scanning electron microscope (FEI Co., Ltd., Portland, OR, USA) to distinguish silica morphology between AS and PS before adding them to the rubber latex. Silica suspensions were dropped on silica wafers and sputter-coated with a thin layer of gold after drying at room temperature, to prevent electrical charging during examination. The measurement was performed at an accelerating voltage of 15 kV and operation distance was 10 mm.

The particle size of the silica suspensions was measured with a dynamic light scattering (DLS) instrument (Mastersizer 2000, Malvern Co., Ltd., Malvern city, UK), corresponding with the characterization by SEM. A trace of emulsifier (OP-10) was of great use to make silica particles discrete during testing.

The difference in reactive groups between pure and modified silica was identified on a Tensor 27 FTIR spectrometer (Bruker Co., Ltd., Ettlingen, Germany). Amounts of 10 mg of pure and modified silica powder were added into the mortar and ground with dried KBr powder; the mass ratio of silica to KBr was 0.0125. Infrared absorption tests were performed at the wavelength range of 400–4000 cm^{-1} and 32 scans were conducted.

Silica weight loss was measured on a TG209 thermo-gravimetric analyzer (NETZSCH Co., Ltd., Selb, Germany) under nitrogen atmosphere. The samples were heated from 30 to 850 °C at a heating rate of 10 °C/min.

The dynamic rheological properties of the silica/ESBR compounds were analyzed on a RPA2000 rubber process analyzer (Alpha Technologies Co., Ltd., Akron, OH, USA) at 60 °C. The strain sweep amplitude varied from 0.2 to 100% at the test frequency of 1 Hz. The curves of the storage modulus (G′)-strain (ε) were obtained.

The dynamic viscoelastic properties of the silica/ESBR vulcanizates were measured on a TQ800 (TA Co., Ltd., New Castle, DE, USA) in tension mode. The temperature was varied from −80 to 80 °C at a heating rate of 3 °C/min. The test frequency was 10 Hz and the strain amplitude was 0.2%. The loss angle tangent (tanδ) was measured as a function of temperature for samples under identical conditions. Long striped specimens were prepared from dumbbell-shaped specimens and both ends were cut off.

The physical–mechanical properties of the silica/ESBR vulcanizates, such as the tensile properties including the tensile strength, modulus at 300% elongation and elongation at break (ASTM D410), were determined using an Electrical Tensile Tester (AT-7000S, ZWICK Co., Ltd., Ulm-Einsingen, Germany) at a tensile rate of 500 mm/min. The specimens were prepared to a dumbbell shape that was punched out from a molded sheet. Five specimens were measured for each sample, and the average values were calculated and reported.

The silica dispersion in silica/ESBR vulcanizates was observed under a JSM-7500F scanning electron microscope (JEOL Co., Ltd., Shoshima City, Japan) with an accelerating voltage of 5 kV. The vulcanized rubber strips were frozen in liquid nitrogen and broken in their brittle state, then adhered to the conductive adhesive directly with the broken section exposed, and finally sprayed with gold in a vacuum environment.

3. Results

3.1. Characterization of Pure and Modified Silica

3.1.1. SEM and DLS Results of Pure Silica

Based on our previous study [17], the morphology and size of the silica particles added to ESBR latex were measured using SEM and DLS, respectively. As shown in Figure S1, AS showed a morphology of spherical particles with an average size of 28 nm. In contrast, PS exhibited a broad distribution of particle sizes due to severe aggregation.

3.1.2. Silane Coupling Agent Si747

Figure 1a shows the chemical structure of the water-soluble silane coupling agent Si747. The Si atom is connected to a mercaptopropyl, of which the free mercapto group can be regarded as an active point to couple with the rubber chain in the vulcanization process. In addition, a methyoxyl group connected to the Si atom hydrolyzed to form Si–OH in contact with water, then dehydrated and condensed with the Si–OH on the surface of the silica, resulting in being chemically grafted onto silica surface. Subsequently, two long chains of polymeric substituents replace the volatile and easily hydrolysable methyoxyl group, which leads to the reduced emission of VOC. The polymeric substituents contain a polar polyether part and a hydrophobic alkyl part. The polyether part with its polar property ensures high silica affinity and fast adsorption and reaction on the silica surface, which compensate for the steric hindrance effect caused by the excessive volume of the substituent. Meanwhile, the alkyl part derived from olefin polymerization are at the end of the long polymeric substituents. On the one hand, the alkyl can shield the free mercapto group and delay the activating reaction of Si747, resulting in improved anti-scorch performance of silica/rubber compounds; on the other hand, the extra alkyls also shield the silanol group, leading to excellent hydrophobation of the silica. This weakens the interaction between the silica particles and improves dispersion. As depicted in Figure 1b, the long-chain nature of the polymeric substituents provides a special shielding effect.

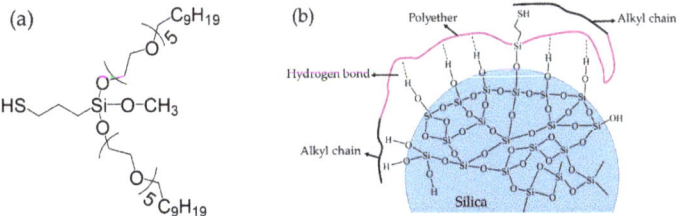

Figure 1. (a) Chemical structure of silane coupling agent Si747 and (b) schematic diagram of Si747 attached to the silica surface.

Figure 2 shows the solution state of Si747 hydrolyzed for 12 h at different pH and a constant temperature of 30 °C. As we all know, there exists Si–OCH$_3$ in the structure of Si747, which is easy to hydrolyze [29,30]. However, the condensation rate of the hydrolysates depends on the acid concentration, oxyhydrogen anion concentration, and the structure around the Si747 hydrolysis groups in the solution system [29]. Therefore, the pH value of the solution is the key parameter for controlling the relative rate and range of the competing process, which is the hydrolysis and condensation of the silane coupling agent [31]. As shown in Figure 2, at neutral pH (pH = 7), the whole hydrolysis solution showed a white turbid state, which was basically maintained as the initial state as Si747 was added into the solution, indicating that hydrolysis and condensation were weak. In a low pH (acid, pH = 3) solution, white floc presented, which was the self-condensate products formed by the hydrolysis and condensation of Si747. By comparison, in an alkaline environment, the whole solution system became transparent under strong alkaline (pH = 12) conditions; under weak alkaline conditions (pH = 9), the top layer of solution presented transparent and the bottom layer remained a white turbid substance, which was the unhydrolyzed Si747. All this indicates that Si747 could hydrolyze under alkaline conditions, and the rate of hydrolysis in a weak alkaline environment was slower than that in a strong alkaline condition [29]. All hydrolysates incurred no obvious condensation under alkaline conditions, and no self-condensate products formed, which may be related to the chemical environment of Si–OCH$_3$ changed via the two long-chain polymeric substituents. Overall, Si747 had high hydrolysis and condensation rate in a strong acid system [31], while the lowest hydrolysis rate was obtained at neutral pH [29]. High hydrolysis and slow condensation rate were obtained under a strong alkaline condition [29], while moderate hydrolysis and condensation rate were obtained under a weak alkaline condition (pH = 9).

Figure 2. Pictures of Si747 hydrolyzed for 12 h at different pH and 30 °C.

3.1.3. FTIR Spectra of Pure and Modified Silica

The FTIR spectra of the two types of silica modified with different dosages of Si747 after rinsing are shown in Figure 3. It could be seen that the absorption peaks at 3440 and 1635 cm^{-1} corresponded to the stretching and deforming vibration modes, respectively, of the H–O–H bonds in the adsorbed water [32]; the absorbance ranging from 1000 to 1150 cm^{-1} was assigned to the Si–O–Si asymmetric stretching mode. Comparing the FTIR spectra of modified silica with pure silica, all modified silica had adsorption peaks at 2925 and 2861 cm^{-1} in the spectra curves, which were attributed to the vibrations of –CH$_2$– and –CH$_3$– bonds [24,33]. These –CH$_2$– and –CH$_3$– bonds that derive from Si747 were detected after rinsing, which proved that Si747 was successfully grafted onto the silica surface.

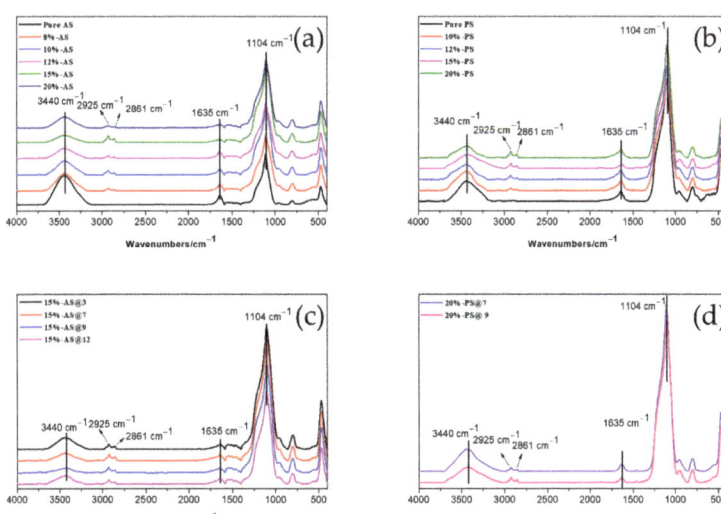

Figure 3. FTIR spectra of pure silica and silica modified with Si747: (**a**) AS; (**b**) PS; (**c**) AS modified at different pH; (**d**) PS modified at different pH.

Figure 3c,d exhibit the FTIR spectra of the two types of silica modified with Si747 at different pH and constant 80 °C after rinsing.

It is well known that the relative intensity (RI) of the peak at 3440 cm^{-1} is determined by the number of –OH bonds [24,33]. Therefore, the higher the RI of the peak at 3440 cm^{-1}, the greater the number of hydroxyl groups on the silica surface. Shown in Table 2 are the RI values, which were normalized to the calculated 3440 cm^{-1} peak in the FTIR spectra of the AS and PS before and after modification, respectively. It can be seen that the RI of the pure silica was higher than that of the silica modified with Si747. This is due to the chemical grafting and physical shielding of the silanol on the silica surface. For the modified AS, as the dosage of Si747 increased from 8% to 20%, the RI of the peak at 3440 cm^{-1} first decreased, then increased. The minimum value was reached when the amount of Si747 was 15%. As the amount of Si747 continued to increase to 20%, the RI value increased inversely. This is because during the hydrolysis process of Si747, self-condensation proceeded at the same time, and the degree of self-condensation increased as the dosage of Si747 increased. These self-condensates could sink into the aqueous system as precipitates, or be physically adsorbed on the silica surface and prevent the hydrolyzed Si747 contacting the silanol, resulting in the increased RI of the peak at 3440 cm+ after the adsorbed Si747 was rinsed off with toluene. For PS, the RI of the peak at 3440 cm^{-1} continued to decrease as the amount of Si747 increased. The minimum value was reached with 20% Si747. This may be because the surface activity of PS was higher than that of AS, and aggregation was more serious. A larger amount of Si747 was required.

Moreover, both the modified AS and PS presented the lowest RI of the peak at 3440 cm^{-1} at pH = 9, indicating that the silanol density on the surface was the lowest and the modification effect was optimal.

Table 2. Relative intensities of pure silica and modified silica at the 3440 cm^{-1} peak.

Sample	Pure AS	8%-AS	10%-AS	12%-AS	15%-AS	20%-AS
RI [1]/cm^{-1}	0.242	0.1432	0.118	0.0844	0.0596	0.0896
Sample	Pure PS	-	10%-PS	12%-PS	15%-PS	20%-PS
RI/cm^{-1}	0.0684	-	0.0653	0.0500	0.0459	0.0404
Sample	15%-AS@3	15%-AS@7	15%-AS@9	15%-AS@12	20%-PS@7	20%-PS@9
RI/cm^{-1}	0.05961	0.03934	0.0243	0.04442	0.40098	0.27502

[1] Relative intensity.

3.1.4. TGA Analysis of Pure and Modified Silica

Figure 4 shows the weight loss of pure and modified silica with different amounts of Si747 and at different pH after rinsing. It can be seen from the figures that over the process of temperature change, the weight loss curves of silica can be divided into two regions. In the first region, where the temperature was below 125 °C, the weight loss on the thermogravimetric curves was mainly due to the loss of adsorbed water on the silica surface. It should be noted that the weight loss in pure silica was higher than that in modified silica in this region, which indicates that the amounts of silanol and adsorbed water on the modified silica surfaces decreased. As the temperature increased in the second region, 125–800 °C, there was greater weight loss relative to the first region. The weight loss in pure silica was mainly the dehydroxylation of silica surface silanol, while that in the modified silica was not only the dehydroxylation of unreacted silanol but also the thermal decomposition of grafted Si747. For modified PS, the weight loss increased as the Si747 dosage increased, and the maximum value was reached with 20% Si747. For AS, weight loss was not very different with different Si747 amounts. However, as the amount of Si747 continued to increase, weight loss first increased, then decreased. Weight loss reached its maximum value with 15% Si747. These results were consistent with the RI pattern in Table 2. In conclusion, modified AS with 15% Si747 and PS with 20% Si747 exhibited the best effects. In addition, comparing the residual mass of the two types of modified silica with the optimum amount of Si747 at 800 °C indicated that PS had a higher degree of Si747 grafting.

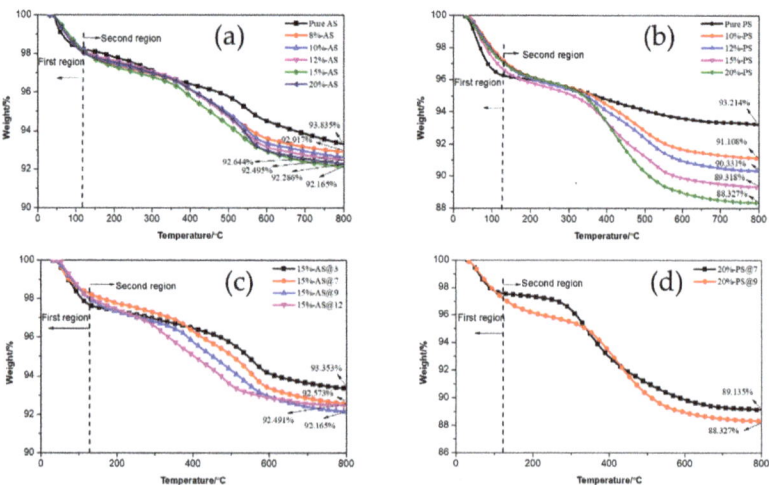

Figure 4. TGA curves of pure silica and silica modified with Si747: (a) AS; (b) PS; (c) AS modified at different pH; (d) PS modified at different pH.

For modified PS, with the increase in pH from 7 to 9, weight loss increased from 10.865% to 11.673%, indicating that the degree of Si747 grafting in modified PS was higher at pH = 9. For modified AS, weight loss was still not very different with different pH. However, it was still observed that as the pH continued to increase, the weight loss of silica first increased and then decreased. The maximum value was reached at pH = 9.

3.2. Characterization of ESBR/Silica Compounds

3.2.1. Vulcanization Properties of ESBR/Silica Compounds

It is known that the vulcanization properties of ESBR/silica compounds determine their prospects for application.

Table 3 shows the vulcanization properties of ESBR compounds filled with different types of silica. The scorch and optimum curing time of ESBR/silica compounds are represented by t_{s2} and t_{90}, respectively. The minimum torque M_L reflects the plasticity of compounds at a given temperature, and the maximum torque M_H reflects the modulus of the vulcanizates; M_H–M_L indicates the maximum crosslink density of the composites. As the amount of Si747 increased, scorch resistance weakened, which was indicated by the shortened t_{s2}; the shortened optimum curing time and increased curing rate indicated that the vulcanization process could be promoted by introducing Si747; the decreased M_L meant improved processability. Given the same amount of Si747, the scorch time of ESBR/MAS compounds was slightly shorter than that of ESBR/MPS compounds, which showed that the scorch resistance of the former was slightly weaker than that of the latter. Meanwhile, the curing rate of the former was slower than that of the latter according to the trend in the CRI values. Moreover, when comparing M_L values, the lower M_L values for the ESBR/MAS compounds indicated better processability. As the amount of Si747 increased, the crosslinking density of ESBR/MAS vulcanizates first increased, then decreased, reaching a maximum with 15% Si747, while, for ESBR/MPS vulcanizates, the crosslinking density continued to increase and reached a maximum with 20% Si747. It is worth noting that the crosslinking densities of the ESBR/MPS vulcanizates were greater than those of the ESBR/MAS vulcanizates for the same amount of Si747.

Table 3. Vulcanization characteristics of ESBR compounds filled with different types of silica.

Sample	Compound Types	t_{s2}/min	t_{90}/min	CRI/min^{-1}	M_L/dN·m	M_H/dN·m	M_H–M_L/dN·m
Pure ESBR	-	1.04	4.66	27.62	0.29	4.13	3.84
ESBR/AS compounds	Pure AS-R	5.57	10.94	18.62	2.96	18.13	15.17
	8%-AS-R	1.77	6.43	21.46	1.85	13.81	11.96
	10%-AS-R	1.49	5.29	26.32	1.75	13.78	12.03
	12%-AS-R	1.43	4.01	38.76	1.72	13.94	12.22
	15%-AS-R	1.14	3.43	43.67	1.70	14.37	12.67
	20%-AS-R	0.91	3.09	45.87	1.51	13.28	11.77
	15%-AS@3-R	2.48	6.30	26.18	2.40	10.63	8.23
	15%-AS@7-R	1.31	3.98	37.45	2.18	11.21	9.03
	15%-AS@9-R	1.14	3.13	43.67	1.70	14.37	12.67
	15%-AS@12-R	1.00	2.85	54.05	2.31	13.48	11.17
ESBR/PS compounds	Pure PS-R	2.09	7.02	20.28	2.83	17.18	14.35
	10%-PS-R	1.59	3.52	51.81	2.18	14.27	12.09
	12%-PS-R	1.45	3.33	53.19	2.04	14.43	12.39
	15%-PS-R	1.19	2.69	66.67	1.94	14.54	12.7
	20%-PS-R	0.93	2.39	68.49	1.68	14.66	12.98
	20%-PS@7-R	1.51	3.72	45.25	1.89	13.02	11.13
	20%-PS@9-R	0.93	2.39	68.49	1.68	14.66	12.98

For ESBR/MAS compounds, as pH increased from acidic to alkaline, anti-scorch ability reduced as t_{s2} decreased. The optimum curing time t_{90} was shortened and the curing rate obviously increased. All these reached their maximum values at pH = 12. It is known that there is a large amount of hydroxyl groups on the silica surface, which causes the silica to be acidic and promotes the adsorption of alkaline accelerators (such as DPG) to delay vulcanization [1]. The AS was modified with Si747 at pH = 3 and blended with ESBR latex to co-flocculate. The compound system continued to be acidic and was not good for vulcanization. By contrast, as the pH increased, the basicity of the compound system gradually increased, which promoted the vulcanization process, including an increased curing rate and shortened scorch time. It should be noted that stronger alkalinity yielded inferior reinforcement of vulcanizates. Meanwhile, as pH increased from acidic to alkaline, the M_L values of the compounds and M_H and M_H–M_L values of the vulcanizates all showed a trend of first decreasing, then increasing. This is because with the increase in pH, the degree of Si747 grafting on AS increased first, then decreased, reaching its maximum at pH = 9; as a result, as the degree of grafting increased, the silanol density on the silica surface decreased, the silica agglomerates weakened, the silica dispersion in the rubber matrix improved and the interaction between silica and rubber molecules was enhanced, which resulted in a decrease in the initial modulus M_L of the ESBR/silica compounds and an enhancement in processability. After vulcanization, all final moduli M_H and modulus differences M_H–M_L in the vulcanizates increased. When pH continued to increase to 12, the grafting degree of Si747 on AS decreased, the particle agglomeration intensified, the M_L of compounds increased and plasticity was inferior as a result. Meanwhile, the modulus difference M_H–M_L, indicating the crosslinking density of vulcanizates, decreased.

For PS modified with Si747 at different pH values then filled with ESBR, the pattern of change in vulcanization characteristics was similar to AS.

3.2.2. Dynamic Mechanical Properties of ESBR/Silica Compounds

Figure 5 exhibits the G′-Strain curves of ESBR compounds filled with different types of silica modified with Si747. Tables S1–S3 show the shear storage modulus difference ΔG′ of ESBR compounds filled with AS and PS before and after modification, respectively.

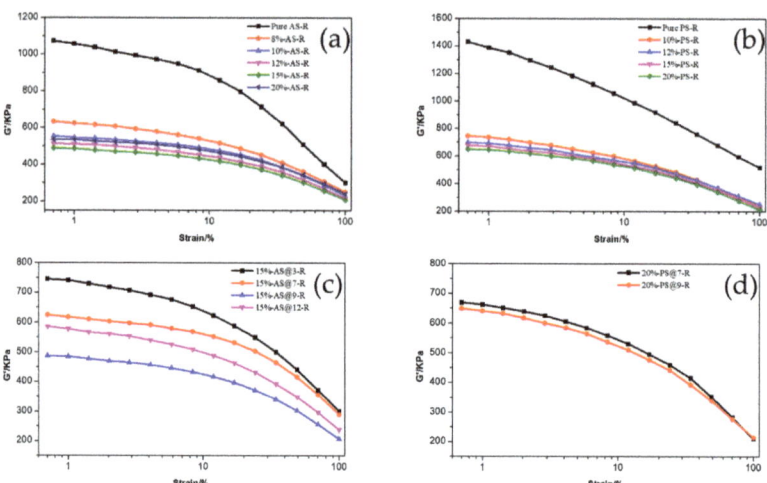

Figure 5. G′-Strain curves of ESBR compounds filled with different types of silica: (**a**) AS; (**b**) PS; (**c**) AS modified at different pH; (**d**) PS modified at different pH.

In general, the phenomenon in which the shear storage modulus G′ of silica-filled rubber drops sharply with increasing strain is called the Payne effect. The difference in shear storage G′ between low strain and high strain can indicate the strength of the Payne effect, which reveals the dispersibility of the silica. The smaller the ΔG′, the weaker the Payne effect, and the better the dispersibility of the silica [34]. As shown in Figure 5, the shear storage modulus G′ and modulus difference ΔG′ of ESBR compounds filled with modified silica were lower than that of silica before modification and filling. The Payne effect of ESBR compounds filled with silica modified with Si747 was weakened and the dispersion of silica was improved. Meanwhile, the shear storage modulus difference

ΔG′ of ESBR/AS compounds was always lower than that of ESBR/PS compounds in the corresponding state, which indicated that ESBR/AS compounds presented with weaker filler networks and Payne effects. The interactions between silica–silica particles were weak and the dispersion was good.

For ESBR/MAS compounds with AS modified with different amounts of Si747, as the amount of Si747 continued to increase, the degree of modification of monodisperse silica increased, while the initial modulus and ΔG′ of ESBR/MAS compounds gradually decreased, indicating that the Payne effect was weakened and the dispersion of silica was improved. When the amount of Si747 reached 15%, the modification degree of the silica surface was the highest and the initial modulus and ΔG′ of ESBR/MAS compounds were the lowest, indicating the weakest silica network and Payne effect along with the best relative filler dispersion.

For ESBR/MPS compounds, it is worth noting that as the amount of Si747 increased, the initial modulus and ΔG′ of ESBR/MPS compounds decreased continuously, indicating that the Payne effect and silica filler network were weakened and the silica dispersion was increasing. It reached optimal levels when the amount of Si747 was 20% by weight.

In summary, comparing AS with PS before and after modification when used to fill ESBR compounds, at constant amounts of Si747, ESBR/AS compounds had lower filler networks and Payne effects; therefore, AS dispersion was better.

In addition, as pH gradually increased, the initial modulus and ΔG′ of ESBR/MAS compounds with AS modified at different pH showed a trend of first decreasing, then increasing. The minimum values were reached at pH = 9.

It is worth noting that comparing ESBR/MAS compounds with AS modified with 15% Si747 at pH = 9 with PS modified with 20% Si747 at pH = 9, the initial modulus and ΔG′ of the former compounds were smaller than that of the latter.

3.3. Characterization of ESBR/Silica Vulcanizates

3.3.1. Dynamic Viscoelastic Properties of ESBR/Silica Vulcanizates

The dynamic viscoelastic properties of ESBR/silica vulcanizates were tested through dynamic mechanical analysis. Figure 6 shows the temperature dependence of the loss factor of ESBR vulcanizates filled with AS and PS modified with different amounts of Si747. The loss factor (tanδ) first increased and then decreased with increasing temperature (T). The peak of tanδ-T curve represented the glass transition temperature (T_g) for ESBR/silica vulcanizates. In Figure 6a, the ESBR/MAS vulcanizate with AS modified with 15% Si747 had the largest loss factor at T_g; meanwhile, the T_g of this vulcanizate was the highest. Similarly, in Figure 6b, the ESBR/MPS vulcanizate with PS modified with 20% Si747 had the largest loss factor at T_g and the T_g was the highest. As indicated above, AS modified with 15% Si747 and PS modified with 20% Si747 displayed the best dispersion in the ESBR matrix.

The viscoelastic properties of tires, including wet skid resistance and rolling resistance, can be evaluated following the temperature dependence of the loss factor curve when ESBR/silica vulcanizate is applied to the tire tread. High-performance tires require a lower tanδ at 60 °C to reduce rolling resistance, and a higher tanδ at 0 °C to ensure high wet skid resistance. As shown in Figure 6, the ESBR/MAS vulcanizate with AS modified with 15% Si747 had the largest tanδ at 0 °C and the lowest tanδ at 60 °C, indicating that the dynamic viscoelastic properties of this vulcanizate were optimal. Meanwhile, the ESBR/MPS vulcanizate with PS modified with 20% Si747 had the best dynamic viscoelastic properties.

Observing the tanδ values in Tables S4 and S5, the wet skid resistance of ESBR/MAS vulcanizate with AS modified with 15% Si747 was enhanced by 4.27% compared with the ESBR/MPS vulcanizate with PS modified with 20% Si747; meanwhile, the rolling resistance was reduced by 13.92%.

As shown in Figure 6c,d, the vulcanizates exhibited the largest loss at the glass transition and the highest Tg when AS and PS were modified with Si747 at pH = 9, which indicated the silica with the best dispersion and strongest interaction with rubber. In addition, as shown in Table S6, ESBR vulcanizates filled with silica modified with Si747 at pH = 9 had the highest tanδ at 0 °C and lowest tanδ at 60 °C, which indicated that these vulcanizates had the optimal dynamic viscoelastic properties.

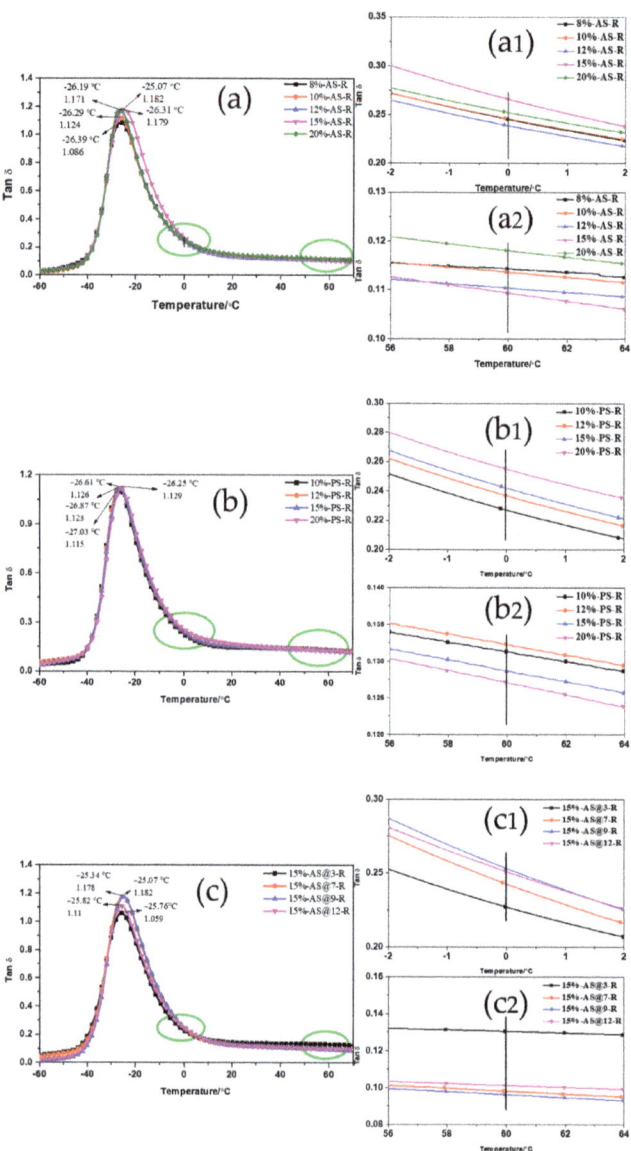

Figure 6. *Cont.*

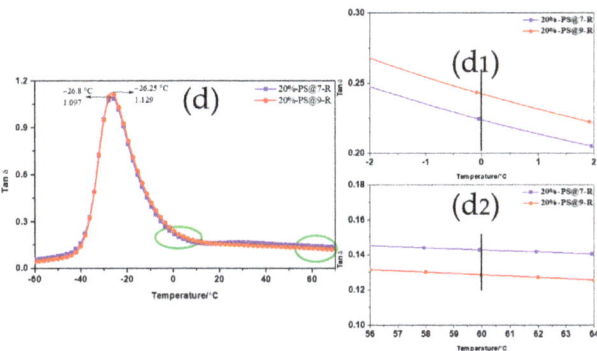

Figure 6. Temperature dependence of the loss factor of ESBR/MAS and ESBR/MPS vulcanizates: (**a**) AS; (**a1**) The tanδ at 0 °C; (**a2**) The tanδ at 60 °C; (**b**) PS; (**b1**) The tanδ at 0 °C; (**b2**) The tanδ at 60 °C; (**c**) AS modified at different pH; (**c1**) The tanδ at 0 °C; (**c2**) The tanδ at 60 °C; (**d**) PS modified at different pH; (**d1**) The tanδ at 0 °C; (**d2**) The tanδ at 60 °C.

3.3.2. Physical Mechanical Properties of ESBR/Silica Vulcanizates

As shown in Table 4, ESRB/MAS vulcanizate with AS modified with 15% Si747 exhibited the highest tensile strength, modulus at 300% elongation, reinforcing index and elongation at break. This performance in the mechanical properties of the vulcanizates was also due to the dispersion of silica, the interaction between silica and the rubber matrix, and the crosslinking density. Si747 could react with rubber through the mercaptopropyl group, thereby forming a "coupling bridge" between silica and rubber. The reinforcing efficiency of silica on rubber was enhanced via the increase of chemical interaction between silica and rubber [24]. Based on the data, AS modified with 15% Si747 had the maximum grafting degree (Figure 4); the modified AS-filled ESBR compounds had the best dispersion of silica and weakest Payne effect (Figure 5); the vulcanizate had the highest crosslinking density (Table 3). All of these contributed to reinforcing the rubber. Similarly, the ESBR/MPS compound with PS modified with 20% Si747 also presented the highest tensile strength, modulus at 300% elongation, reinforcing index and elongation at break.

Table 4. Physical mechanical properties of ESBR/AS vulcanizates before and after modification.

Sample	Pure ESBR	Pure AS-R	8%-AS-R	10%-AS-R	12%-AS-R	15%-AS-R	20%-AS-R
Shore A/°	41	63	56	55	54	53	54
Tensile strength/MPa	1.90 ± 0.05	4.46 ± 0.06	10.88 ± 0.12	11.67 ± 0.15	12.42 ± 0.22	15.70 ± 0.52	13.33 ± 0.41
Modulus at 100% elongation/MPa	0.63 ± 0.02	1.41 ± 0.04	1.16 ± 0.03	1.11 ± 0.03	0.99 ± 0.02	1.09 ± 0.03	1.10 ± 0.03
Modulus at 300% elongation/MPa	1.13 ± 0.02	2.22 ± 0.05	2.63 ± 0.06	2.78 ± 0.06	2.82 ± 0.07	3.75 ± 0.08	3.64 ± 0.07
Reinforcing index	1.79	1.57	2.27	2.50	2.91	3.44	3.31
Elongation at break/%	520 ± 13	476 ± 12	583 ± 15	672 ± 22	687 ± 23	716 ± 28	664 ± 21

It is noted that, according to Tables 4 and 5, the ESBR/MAS vulcanizate exhibited lower hardness, modulus at 100% and 300% elongation, and reinforcing index compared to the ESBR/MPS vulcanizate at the same amount of Si747. All these may be related to the lower crosslinking density of the former compared to that of the latter, while the tensile strength and elongation at break of the former had obvious advantages over the latter.

The tensile strength of ESBR/MAS vulcanizates with AS modified with 15% Si747 was enhanced by 14.8% compared to that of ESBR/MPS vulcanizates with PS modified with 20% Si747; meanwhile, elongation at break was enhanced by 62.4%, and hardness and modulus at 300% elongation were decreased by 13.1% and 53.2%, respectively.

Table 5. Physical mechanical properties of ESBR/PS vulcanizates before and after modification.

Sample	Pure ESBR	Pure PS-R	10%-PS-R	12%-PS-R	15%-PS-R	20%-PS-R
Shore A/°	41	68	61	59	60	61
Tensile strength/MPa	1.90 ± 0.05	8.57 ± 0.08	10.69 ± 0.10	11.45 ± 0.15	11.95 ± 0.15	13.68 ± 0.20
Modulus at 100% elongation/MPa	0.63 ± 0.02	3.49 ± 0.06	1.55 ± 0.03	1.39 ± 0.04	1.41 ± 0.04	1.43 ± 0.03
Modulus at 300% elongation/MPa	1.13 ± 0.02	-	6.23 ± 0.08	7.02 ± 0.08	7.55 ± 0.08	8.20 ± 0.09
Reinforcing index	1.79	-	4.02	5.05	5.35	5.73
Elongation at break/%	520 ± 13	271 ± 8	374 ± 9	380 ± 9	438 ± 10	441 ± 10

The physical mechanical properties of the ESBR vulcanizates filled with two types of silica modified at different pH are shown in Table 6. For AS, with the increase in pH from acidic to alkaline, the tensile strength of the ESBR/silica vulcanizates obviously first increased, then decreased, reaching its maximum at pH = 9. Likewise, the modulus at 300% elongation, the reinforcing index and elongation at break of vulcanizates all presented the same tendency as the tensile strength as pH increased. These are mainly related to the chemical interaction between silica and rubber; that is, the crosslinking density of vulcanizates played an important role. At pH = 9, AS with the highest grafting degree of Si747, which can combine silica with rubber via chemical reaction, dispersed well in the rubber matrix and had the strongest interaction with the rubber molecule chain. In conclusion, the ESBR/MAS vulcanizates had excellent physical mechanical properties with the highest crosslinking density.

Table 6. Physical mechanical properties of ESBR vulcanizates filled with AS and PS modified with Si747 at different pH.

Sample	15%-AS@3-R	15%-AS@7-R	15%-AS@9-R	15%-AS@12-R	20%-PS@7-R	20%-PS@9-R
Shore A/°	60	54	53	57	62	61
Tensile strength/MPa	11.47 ± 0.12	13.06 ± 0.22	15.70 ± 0.52	13.90 ± 0.25	12.76 ± 0.20	13.68 ± 0.20
Modulus at 100% elongation/MPa	1.46 ± 0.03	1.21 ± 0.03	1.09 ± 0.03	1.10 ± 0.03	1.54 ± 0.04	1.43 ± 0.03
Modulus at 300% elongation/MPa	3.30 ± 0.06	3.60 ± 0.05	3.75 ± 0.08	3.15 ± 0.06	7.02 ± 0.08	8.20 ± 0.09
Reinforcing index	2.26	2.98	3.44	2.92	4.56	5.73
Elongation at break/%	496 ± 15	630 ± 25	716 ± 28	608 ± 23	434 ± 12	441 ± 10

In contrast, the modulus at 100% elongation also reflected the ability of the vulcanizates to resist external force without deformation and exhibited a change tendency opposite from that of the modulus at 300% elongation. At this time, due to the low deformation rate of the vulcanizates, the filler network was the main contribution to the modulus. As modification efficiency increased, the silica dispersed better in the rubber, and the modulus at 100% elongation of vulcanizates decreased.

Compared to those at neutral pH, ESBR/MPS vulcanizates with PS modified at pH = 9 exhibited better physical mechanical properties, such as tensile strength, modulus at 300% elongation, reinforcing index and elongation at break.

3.3.3. Micromorphology of ESBR/Silica Vulcanizates

Figures 7 and 8 show SEM images of the brittle fracture section of ESBR vulcanizates filled with different types of silica before and after modification. The bare white particles in the images are the silica under contrast via SEM. The strength of the interaction between silica and rubber can be determined by observing the amount of bare silica. In the process of brittle fracture, rubber molecules slipped off the surface of the silica and the fracture site should be a cross-linked network if the silica particles interacted strongly with the rubber molecules. If a rubber molecule separated from the silica surface, the silica would be exposed and distributed at the fracture section. As well, the more homogeneous the dispersion of silica, the better the comprehensive performance of the ESBR vulcanizate [23].

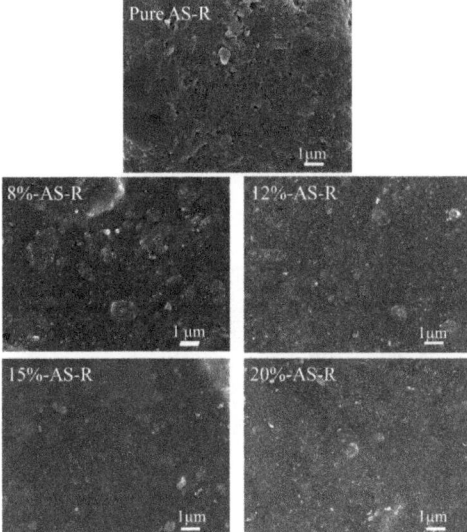

Figure 7. Fracture surface SEM micrographs of ESBR/AS vulcanizates before and after modification.

It is known that, as the grafting degree of Si747 increased, the dimension of silica aggregates decreased and the modified silica dispersed more uniformly in the rubber, enhancing the interaction between modified silica and rubber. As shown in Figure 7, as the amount of Si747 increased from 8% to 20%, the aggregation of silica in rubber matrix first weakened, then slightly strengthened. Meanwhile, the exposed amount of silica first decreased, then increased, and the minimum value was reached when 15% Si747 was used, where the silica interacted with the rubber molecules most strongly. Similarly, as shown in Figure 8, for the ESBR/PS vulcanizates, when 20% Si747 was used, the amount of exposed silica in the rubber matrix was minimal and silica was dispersed most uniformly.

Figures 9 and S2 show SEM images of the brittle fracture sections of ESBR vulcanizates filled with AS and PS modified at different pH, respectively. In Figure 9, large AS aggregates can be observed clearly in the rubber matrix at pH = 3 and 7, implying inefficient modification and inferior dispersion of the silica. At a pH of 9 (shown in Figure 7 15%-AS-R), the vulcanizate had fewer silica aggregates and bare silica particles in comparison to the others. As the pH continuously increased above 9, more aggregates and bare particles appeared, indicating weaker interactions with rubber molecules.

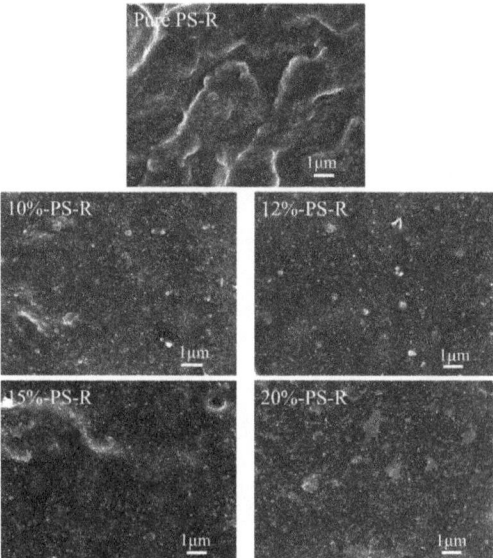

Figure 8. Fracture surface SEM micrographs of ESBR/PS vulcanizates before and after modification.

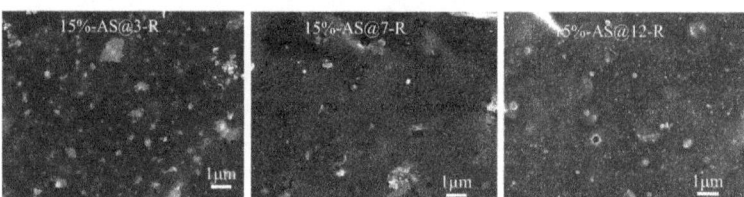

Figure 9. Fracture surface SEM micrographs of ESBR/MAS vulcanizates with AS modified at different pH.

For PS (Figures S2 and 8 20%-PS-R), as the modification pH increased from 7 to 9, a small portion of rubber film existed at the fracture surface of the vulcanizate, implying that the interaction between the silica and the rubber matrix became stronger.

4. Conclusions

In this study, AS was successfully synthesized through the hydrolysis and condensation of tetraethoxysilane with L-lysine as the catalyst; the size and surface area of AS could be controlled similar to that of PS. Both AS and PS slurry could be modified using water-soluble 3-mercaptopropyloxy-methoxyl-bis(nonane-pentaethoxy) siloxane (Si747) directly and then blended with ESBR latex to prepare silica/ESBR master batches via the latex compounding technique. In this article, the effect of amount of Si747 and modification pH on silica and the comprehensive properties of the resulting silica/ESBR composites were systematically examined. The modification of silica was achieved by chemical condensation between hydrolyzed Si747 and silanol on the silica surface, along with physical absorption between a polar polyether in two long-chain polymeric substituents and silanol on the silica surface. FT-IR and TGA results for modified AS or PS powder showed that the optimum amounts of Si747 for AS and PS were 15% and 20%, respectively. Meanwhile, the optimum modification pH was 9. ESBR/MAS compounds presented weaker filler networks and stronger filler–rubber interactions than ESBR/MPS compounds with the silica modified under the optimum modification condition. Comparing the physical mechanical properties of the ESBR/MAS and ESBR/MPS vulcanizates, the tensile strength and elongation at break were enhanced by 14.8% and 62.4%, respectively, while the hardness and modulus at 300% elongation were decreased by 13.1% and 53.2%, respectively. Notably, for the dynamic properties, the ESBR/MAS vulcanizate exhibited a better combination of low rolling resistance with high wet skid resistance. Considering the performance of monodisperse silica filled ESBR composites, AS may be applied as an ideal substitution for PS in the green tire industry.

Supplementary Materials: The following supporting information can be downloaded at: https://www.mdpi.com/article/10.3390/polym15040981/s1. Figure S1. SEM images and corresponding DLS results of silica particles: (a) AS; (b) corresponding DLS results of AS with single peak and average size of 28 nm; (c) PS; and (d) corresponding DLS results of PS with multiple peaks and average size of 370 nm. Figure S2. Fracture surface SEM micrographs of ESBR/PS vulcanizates with PS modified at different pH. Table S1. Shear storage modulus difference $\Delta G'$ of ESBR compounds filled with monodisperse silica before and after modification. Table S2. Shear storage modulus difference $\Delta G'$ of ESBR compounds filled with precipitated silica before and after modification. Table S3. Shear storage modulus difference $\Delta G'$ of ESBR compounds filled with silica modified at different pH. Table S4. Tanδ at 0 °C and 60 °C of ESBR vulcanizates filled with monodisperse silica modified with different dosages of Si747. Table S5. Tanδ at 0 °C and 60 °C of ESBR vulcanizates filled with precipitated silica modified with different dosages of Si747. Table S6. Shear storage modulus difference $\Delta G'$ of ESBR compounds filled with silica modified at different pH.

Author Contributions: Conceptualization, Z.K. and S.L.; methodology, Z.K.; software, L.X.; validation, L.X., A.T. and J.C.; formal analysis, L.X. and Z.K.; investigation, L.X., A.T. and J.C.; resources, L.X. and Z.K.; data curation, L.X., A.T. and J.C.; writing—original draft preparation, L.X., A.T., J.C. and A.S.; writing—review and editing, Z.K. and S.L.; supervision, Z.K.; project administration, Z.K.; funding acquisition, Z.K. and S.L. All authors have read and agreed to the published version of the manuscript.

Funding: This research was funded by the National Natural Science Foundation of China, grant number 51803104 and the Department of Science and Technology of Shandong Province, grant number 2021CXGC010902.

Institutional Review Board Statement: Not applicable.

Informed Consent Statement: Not applicable.

Data Availability Statement: Not applicable.

Conflicts of Interest: The authors declare no conflict of interest.

References

1. Sengloyluan, K.; Sahakaro, K.; Dierkes, W.K.; Noordermeer, J.W.M. Silica-reinforced tire tread compounds compatibilized by using epoxidized natural rubber. *Eur. Polym. J.* **2014**, *51*, 69–79. [CrossRef]
2. Prasertsri, S.; Rattanasom, N. Fumed and precipitated silica reinforced natural rubber composites prepared from latex system: Mechanical and dynamic properties. *Polym. Test.* **2012**, *31*, 593–605. [CrossRef]
3. Sarkawi, S.S.; Dierkes, W.K.; Noordermeer, J.W.M. Elucidation of filler-to-filler and filler-to-rubber interactions in silica-reinforced natural rubber by TEM Network Visualization. *Eur. Polym. J.* **2014**, *54*, 118–127. [CrossRef]
4. Bhattacharyya, S.; Lodha, V.; Dasgupta, S.; Mukhopadhyay, R.; Guha, A.; Sarkar, P.; Saha, T.; Bhowmick, A.K. Influence of highly dispersible silica filler on the physical properties, tearing energy, and abrasion resistance of tire tread compound. *J. Appl. Polym. Sci.* **2019**, *136*, 47560. [CrossRef]
5. Sun, C.Z.; Wen, S.P.; Ma, H.W.; Li, Y.; Chen, L.; Wang, Z.; Yuan, B.B.; Liu, L. Improvement of Silica Dispersion in Solution Polymerized Styrene–Butadiene Rubber via Introducing Amino Functional Groups. *Ind. Eng. Chem. Res.* **2019**, *58*, 1454–1461. [CrossRef]
6. Kaewsakul, W.; Sahakaro, K.; Dierkes, W.K.; Noordermeer, J.W.M. Optimization of mixing conditions for silica-reinforced natural rubber tire tread compounds. *Rubber Chem. Technol.* **2012**, *85*, 277–294. [CrossRef]

7. Qu, L.L.; Wang, L.J.; Xie, X.M.; Yu, G.Z.; Bu, S.H. Contribution of silica–rubber interactions on the viscoelastic behaviors of modified solution polymerized styrene butadiene rubbers (M-S-SBRs) filled with silica. *RSC Adv.* **2014**, *4*, 64354–64363. [CrossRef]
8. Zou, Y.K.; He, J.W.; Tang, Z.H.; Zhu, L.X.; Luo, Y.F.; Liu, F. Effect of multifunctional samarium lysine dithiocarbamate on curing properties, static and dynamic mechanical properties of SBR/silica composites. *RSC Adv.* **2016**, *6*, 269–280. [CrossRef]
9. Ye, N.; Zheng, J.C.; Ye, X.; Xue, J.J.; Han, D.L.; Xu, H.S.; Wang, Z.; Zhang, L.Q. Performance enhancement of rubber composites using VOC-Free interfacial silica coupling agent. *Compos. Part B* **2020**, *202*, 108301. [CrossRef]
10. Tian, Q.F.; Zhang, C.H.; Tang, Y.; Liu, Y.L.; Niu, L.Y.; Ding, T.; Li, X.H.; Zhang, Z.J. Preparation of hexamethyl disilazane-surface functionalized nano-silica by controlling surface chemistry and its "agglomeration-collapse" behavior in solution polymerized styrene butadiene rubber/butadiene rubber composites. *Compos. Sci. Technol.* **2021**, *201*, 108482. [CrossRef]
11. Yang, J.S.; Xian, B.; Li, H.X.; Zhang, L.Q.; Han, D.L. Preparation of silica/natural rubber masterbatch using solution compounding. *Polymer* **2022**, *244*, 124661. [CrossRef]
12. Li, Y.; Han, B.Y.; Liu, L.; Zhang, F.Z.; Zhang, L.Q.; Wen, S.P.; Lu, Y.H.; Yang, H.B.; Shen, J. Surface modification of silica by two-step method and properties of solution styrene butadiene rubber (SSBR) nanocomposites filled with modified silica. *Compos. Sci. Technol.* **2013**, *88*, 69–75. [CrossRef]
13. Wahba, L.; D'Arienzo, M.; Donetti, R.; Hanel, T.; Scotti, R.; Tadiello, L.; Morazzoni, F. In situ sol–gel obtained silica–rubber nanocomposites: Influence of the filler precursors on the improvement of the mechanical properties. *RSC Adv.* **2013**, *3*, 5832–5844. [CrossRef]
14. Sittiphan, T.; Prasassarakich, P.; Poompradub, S. Styrene grafted natural rubber reinforced by in situ silica generated via sol–gel technique. *Mater. Sci. Eng. B* **2014**, *181*, 39–45. [CrossRef]
15. Poompradub, S.; Thirakulrati, M.; Prasassarakich, P. In situ generated silica in natural rubber latex via the sol–gel technique and properties of the silica rubber composites. *Mater. Chem. Phy.* **2014**, *144*, 122–131. [CrossRef]
16. Prasertsri, S.; Rattanasom, N. Mechanical and damping properties of silica/natural rubber composites prepared from latex system. *Polym. Test.* **2011**, *30*, 515–526. [CrossRef]
17. Xia, L.J.; Song, J.H.; Wang, H.; Kan, Z. Silica nanoparticles reinforced natural rubber latex composites: The effects of silica dimension and polydispersity on performance. *J. Appl. Polym. Sci.* **2019**, *136*, 1–11. [CrossRef]
18. Ngeow, Y.W.; Heng, J.Y.Y.; Williams, D.R.; Davies, R.T.; Lawrence, K.M.E.; Chapman, A.V. TEM observation of silane coupling agent in silica-filled rubber tyre compound. *J. Rubb. Res.* **2019**, *22*, 1–12. [CrossRef]
19. Sae-oui, P.; Sirisinha, C.; Hatthapanit, K.; Thepsuwan, U. Comparison of reinforcing efficiency between Si-69 and Si-264 in an efficient vulcanization system. *Polym. Test.* **2005**, *24*, 439–446. [CrossRef]
20. Ahn, B.; Kim, D.; Kim, K.; Kim, I.J.; Kim, H.J.; Kang, C.H.; Lee, J.-Y.; Kim, W. Effect of the functional group of silanes on the modification of silica surface and the physical properties of solution styrene-butadiene rubber/silica composites. *Compos. Interface* **2019**, *26*, 585–596. [CrossRef]
21. Luginsland, H.D.; Röben, C. The Development of Sulphur-Functional Silanes as Coupling Agents in Silica-Reinforced Rubber Compounds. Their Historical Development over Several Decades. *Int. Polym. Sci. Tech.* **2016**, *43*, 734–737. [CrossRef]
22. Gui, Y.; Zheng, J.C.; Ye, X.; Han, D.L.; Xi, M.M.; Zhang, L.Q. Preparation and performance of silica/SBR masterbatches with high silica loading by latex compounding method. *Compos. Part B-Eng.* **2016**, *85*, 130–139. [CrossRef]
23. Li, Y.; Han, B.Y.; Wen, S.P.; Lu, Y.L.; Yang, H.B.; Zhang, L.Q.; Liu, L. Effect of the temperature on surface modification of silica and properties of modified silica filled rubber composites. *Compos. Part A-Appl. S* **2014**, *62*, 52–59. [CrossRef]
24. Zheng, J.C.; Han, D.L.; Ye, X.; Wu, X.H.; Wu, Y.P.; Wang, Y.Q.; Zhang, L.Q. Chemical and physical interaction between silane coupling agent with long arms and silica and its effect on silica/natural rubber composites. *Polymer* **2018**, *135*, 200–210. [CrossRef]
25. Bel-Hassen, R.; Boufi, S.; Salon, M.-C.B.; Abdelmouleh, M.; Belgacem, M.N. Adsorption of silane onto cellulose fibers. II. The effect of pH on silane hydrolysis, condensation, and adsorption behavior. *J. Appl. Polym. Sci.* **2008**, *108*, 1958–1968. [CrossRef]
26. Pantoja, M.; Velasco, F.; Broekema, D.; Abenojar, J.; del Real, J.C. The Influence of pH on the Hydrolysis Process of γ-Methacryloxypropyltrimethoxysilane, Analyzed by FT-IR, and the Silanization of Electrogalvanized Steel. *J. Adhes. Sci. Technol.* **2010**, *24*, 1131–1143. [CrossRef]
27. Rostami, M.; Mohseni, M.; Ranjbar, Z. Investigating the effect of pH on the surface chemistry of an amino silane treated nano silica. *Pigm. Resin Technol.* **2011**, *40*, 363–373. [CrossRef]
28. Yokoi, T.; Wakabayashi, J.; Otsuka, Y.; Fan, W.; Iwama, M.; Watanabe, R.; Aramaki, K.; Shimojima, A.; Tatsumi, T.; Okubo, T. Mechanism of Formation of Uniform-Sized Silica Nanospheres Catalyzed by Basic Amino Acids. *Chem. Mater.* **2009**, *21*, 3719–3729.
29. Osterholtz, F.D.; Pohl, E.R. Kinetics of the hydrolysis and condensation of organofunctional alkoxysilanes: A review. *J. Adhes. Sci. Technol.* **1992**, *6*, 127–149. [CrossRef]
30. Beari, F.; Brand, M.; Jenkner, P.; Lehnert, R.; Metternich, H.J.; Monkiewicz, J.; Siesler, H.W. Organofunctional alkoxysilanes in dilute aqueous solution: New accounts on the dynamic structural mutability. *J. Organom. Chem.* **2001**, *625*, 208–216. [CrossRef]
31. Daniels, M.W.; Sefcik, J.; Lorraine, F.; Francis, L.F.; McCormick, A.V. Reactions of a Trifunctional Silane Coupling Agent in the Presence of Colloidal Silica Sols in Polar Media. *J. Colloid Interf. Sci.* **1999**, *219*, 351–356.
32. Yoshida, T.; Tanabe, T.; Hirano, M.; Muto, S. FT-IR study on the effect of oh content on the damage process in silica glasses irradiated by hydrogen. *Nucl. Instrum. Methods Phys. Res.* **2004**, *218*, 202–208. [CrossRef]

33. Zheng, J.C.; Ye, X.; Han, D.L.; Zhao, S.H.; Wu, X.H.; Wu, Y.P.; Dong, D.; Wang, Y.P.; Zhang, L.Q. Silica Modified by Alcohol Polyoxyethylene Ether and Silane Coupling Agent Together to Achieve High Performance Rubber Composites Using the Latex Compounding Method. *Polymers* **2018**, *10*, 1. [CrossRef] [PubMed]
34. Zhou, H.M.; Song, L.X.; Lu, A.; Jiang, T.; Yu, F.M.; Wang, X.C. Influence of immobilized rubber on the non-linear viscoelasticity of filled silicone rubber with different interfacial interaction of silica. *RSC Adv.* **2016**, *6*, 15155–15166. [CrossRef]

Disclaimer/Publisher's Note: The statements, opinions and data contained in all publications are solely those of the individual author(s) and contributor(s) and not of MDPI and/or the editor(s). MDPI and/or the editor(s) disclaim responsibility for any injury to people or property resulting from any ideas, methods, instructions or products referred to in the content.

Article

The Study of Enteromorpha-Based Reinforcing-Type Flame Retardant on Flame Retardancy and Smoke Suppression of EPDM

Peipei Sun [1], Ziwen Zhou [2], Licong Jiang [2], Shuai Zhao [2,*] and Lin Li [2,*]

[1] Advanced Materials Institute, State Key Laboratory of Biobased Material and Green Papermaking, Qilu University of Technology (Shandong Academy of Sciences), Jinan 250014, China
[2] Key Lab of Rubber-Plastics, Ministry of Education/Shandong Provincial Key Lab of Rubber-Plastics, School of Polymer Science and Engineering, Qingdao University of Science and Technology, Qingdao 266042, China
* Correspondence: zhaoshuai@qust.edu.cn (S.Z.); qustlilin@qust.edu.cn (L.L.)

Abstract: Enteromorpha, as a waste from marine pollution, brings great pressure to environmental governance every year, especially for China. Under the premise of a shortage of industrial materials, taking appropriate measures can turn waste into wealth, which will benefit us a lot. In this work, a bio-based reinforcing-type flame retardant based on Enteromorpha is designed. The designed Enteromorpha-based flame retardant system (AEG) mainly focuses on the reinforcing and flame retardant effects on ethylene-propylene-diene tripolymer (EPDM). For the AEG system, ammonium polyphosphate (APP) serves as both the acid source and the gas source; the simple hybrid material (GN) produced by loading graphene (GE) and Enteromorpha (EN) using tannic acid (TA) as a regulator serves as an acid source and a carbonizing source. The results show that when 40 phr AEG is added, the LOI of EPDM/AEG40 reaches 32.5% and the UL-94 reaches the V-0 level. The PHRR and THR values of EPDM/AEG40 are 325.9 kW/m^2 and 117.6 MJ/m^2, respectively, with decrements of 67.3% and 29.7%, respectively, compared with the results of neat EPDM composite. Especially, the TSP and TSR values of EPDM/AEG40 are reduced from 15.2 m^2 of neat EPDM to 9.9 m^2 with a decrement of 34.9% and reduced from 1715.2 m^2/m^2 of neat EPDM to 1124.5 m^2/m^2 with a decrement of 34.4%, indicating that AEG is effective in flame retardancy and smoke suppression. Meanwhile, the tensile strength and tear strength of EPDM/AEG composites are much higher than neat EPDM, therefore, with the future development of innovate reinforcing-type flame-retardant Enteromorpha, the application of Enteromorpha in the polymer flame-retardant field will surely usher in bright development.

Keywords: Enteromorpha; reinforcing; flame retardant

Citation: Sun, P.; Zhou, Z.; Jiang, L.; Zhao, S.; Li, L. The Study of Enteromorpha-Based Reinforcing-Type Flame Retardant on Flame Retardancy and Smoke Suppression of EPDM. *Polymers* **2023**, *15*, 55. https://doi.org/10.3390/polym15010055

Academic Editors: Yuwei Chen and Yumin Xia

Received: 18 November 2022
Revised: 3 December 2022
Accepted: 6 December 2022
Published: 23 December 2022

Copyright: © 2022 by the authors. Licensee MDPI, Basel, Switzerland. This article is an open access article distributed under the terms and conditions of the Creative Commons Attribution (CC BY) license (https://creativecommons.org/licenses/by/4.0/).

1. Introduction

In recent years, with the prices of raw materials rising, the search for alternative materials from nature has become increasingly serious, especially for the rubber industry. Rubber, as an important commercial material, is inseparable from our lives. For engineering applications, adding fillers to rubber is necessary to improve its modulus, strength, wear resistance and fatigue resistance and other properties. Since 2008, due to the high nitrogen levels caused by fertilizers, the suddenly immense outbreak of Enteromorpha as a kind of green algae in the east coast of China caused severe environmental problems and has threatened coastal aquaculture [1–3]. Qingdao is a beautiful city in China where every year a large amount of Enteromorpha need to be cleaned off the coast to maintain normal shipping, safe fishing and clean swimming. In 2015, more than 70,000 tons of seaweed were treated [4]. However, everything has two sides. Enteromorpha mainly contains cellulose and polysaccharides (43.4–60.2%), crude protein (9.0–14.0%), ether extract (2.0–3.6%), ash (32.0–36.0%), as well as some fatty acid [5,6]. As an important phytochemical of green algae, Enteromorpha prolifera polysaccharides have been extensively studied in explosive growth and have been proved to have various physiological and biological activities, including anti-oxidant [7], immunomodulation [8], anti-bacteria [9], anti-hyperlipidemia [10], anti-tumor [11], anti-cancer [12], anti-viral [13], and anti-coagulant [14] properties; it also regulates gut microbiota [15]. As a researcher in the field of composite materials, Enteromorpha's thin, silk-like microstructure and malleable carbonized performance have been of extreme interest to me [16]. In the rubber industry, fiber with a certain aspect ratio is usually used to improve the dimensional stability of unvulcanized rubber and the comprehensive properties of vulcanized rubber by increasing the green strength of the rubber matrix [17]. Fiber with a proper orientation can change the tensile strength of rubber considerably so that, at low elongation, it is impossible to achieve a sharp rise in stress with particle filler [18–20]. This feature is very useful if large deformation is undesirable [17,21]. Enteromorpha, with its thin, silk-like microstructure, may have the potential to replace traditional rubber-reinforcing fillers, alleviating the serious problem of material shortages. Nevertheless, the malleable carbonized performance of Enteromorpha with typical flame-retardant elements (phosphorus and nitrogen) gives it a certain application potential in the field of flame retardants. As is known to all, the mechanism of flame retardancy can be divided into gas phase and condensed phase flame retardants [22]. More precisely, the main mechanism of condensed phase flame retardants is promotion of the formation of char residues that protect the underlying matrix from further combustion [6]. In our previous work, an intumescent flame-retardant system was successfully designed by applying tannic-acid-functionalized graphene combined with ammonium polyphosphate. The synergistic effects of the designed intumescent flame retardant on the flame retardancy and smoke suppression performances of natural rubber resulted in very satisfactory performances.

In this work, we design another efficient bio-based reinforcement-type intumescent flame-retardant system (AEG) based on Enteromorpha, which is environmentally friendly and has potential for introduction into the polymer flame retardant field. In the AEG system, ammonium polyphosphate (APP) is used as both an acid source and a gas source, and the hybrid material (GN), prepared by GE and EN using TA as a regulator, is the acid source and the carbon source. AEG flame retardant system is mainly focused on the enhancement and flame retardancy effects on ethylene-propylene-diene rubber (EPDM). The microstructure and thermal degradation of EN and AEG before and after combustion are analyzed. The thermal degradation and combustion properties of EPDM/AEG composites were investigated by thermogravimetric and microscale combustion calorimetric analysis, Limited Oxygen Index and UL-94 vertical burning test. The microscopic quality of intumescent char is examined using SEM-EDS, FTIR and Raman. Based on the analysis of thermal decomposition behavior of polymer composites and the characteristics of carbon, the flame-retardant mechanism was put forward.

2. Experimental and Materials

2.1. Materials

Ammonium polyphosphate (APP, type is TF-201 with 1500 polymerization degree) was generously supplied by Taifeng New Flame Retardant Co., Ltd., Deyang, China. Graphene (GE) was supplied by the Sixth Element (Changzhou, China) Material Technology Co., Ltd. (Changzhou, China). Enteromorpha was sourced from the southeast coast of Shandong Peninsula in 2017. Tannic acid (TA, analytical grade) was purchased from Aladdin Technology Co., Ltd. (Shanghai, China). EPDM (8550C, the third monomer content is 5%) was purchased from Alangxingke High Performance Elastomer Co., Ltd. (Shanghai, China). Other reagents, including sulfur, 2-mercaptobenzothiazole (M), Tetramethylthiuram disulfide (TT), zinc oxide (ZnO) and Stearic acid (SA), were all industrial grade and kindly supplied by Qingdao Topsen Chemical Co., Ltd. (Qingdao, China).

2.2. Characterization

Raman spectra analyses were recorded using a Bruker FRS-100S with high-resolution and a CCD detector in the wavelength range of 600 to 2500 cm^{-1}. A Bruker Vertex 70 Fourier Transform Infrared Spectrometer was used to characterize the infrared spectra of modified fillers. JSM-6700F SEM-EDX (Electronics Corp. Tokyo, Japan) was used to characterize the morphology of the residue char, for which the samples should be coated with gold for a certain time. Thermal performances were carried out using a TGA-7 instrument (Perkine Elmer Company, Waltham, MA, USA), with a room temperature to 800 °C at 10 °C/min heating rate in N_2 atmosphere. Dynamic mechanical performances were obtained on a NETZSCH DMA242 machine with tensile mode and a temperature increment of 3 °C·min^{-1} at a fixed frequency of 1 Hz and a fixed strain of 0.5%. Limited Oxygen Index was performed on a HC-2 oxygen index meter (Analysis Instrument Company, Jiangning, China) making a reference to ASTM D2863. Vertical Burning Test, a typical combustion test, was performed on a CFZ-1 vertical burning instrument (Analysis Instrument Co., Jiangning, China) according to ASTM D3801-2010. The heat and smoke index were analyzed using the Cone Calorimeter (Fire Testing Technology, West Sussex, UK) exposed to 35 kW/m^2 external heat flux referring to ASTM E1354/ISO 5660. Mechanical property tests, including tensile and tear, were carried out using an AI-7000S Universal Material Tester, for which a dumbbell specimen under 500 mm·min^{-1} tensile speed was employed according to ISO 528:2009. The optimum curing temperature of the EPDM composites was 155 °C determined by an ALPHA MDR2000 UCAN rheometer.

2.3. Preparation of GN

A wall breaker was used to superfine EN for 30 min at room temperature. GE and EN with varied mass ratios were added into a vial containing 100 mL TA aqueous solution with a fixed concentration of TA, shown in Table S1. The mixture was then mixed at high speed for 30 min to prepare GN. The mentioned reaction product was thoroughly filtered and washed with deionized water in order to completely remove unreacted TA. The resulting filter cake was dried at 70 °C for 72 h in the oven. Thus, GN powder was obtained.

2.4. Preparation of Composite Materials

Flame-retardant AEG was prepared by compounding the prepared GN with APP in a certain proportion (Table S1), and EPDM composites were charged in a 200 mL Banbury mixer at 60 rpm rotor speed under 120 °C; the formula is shown in Table 1. The EPDM was firstly fed into the mixer for premixing for 2 min. SA (1 phr), zinc oxide (5 phr) and N330 (40 phr) were then added in sequence, and the compounding ingredients were mixed for another 8 min. Then, the compounds were discharged onto a two-roll mill, and sulfur (1.5 phr), M (0.5 phr) and TT (1.5 phr) at were added 40 °C and mixed for another 5 min to achieve a sheet with a 2 mm thickness. The compound sheets were cut and discharged onto a compression molding machine at 165 °C for a time equal to the optimum cure time. The sheet was then cooled at room temperature and named as EPDM/AEG, as shown in Scheme 1. The sheets were pressed in a compression molding machine at 165 °C for a time equal to the optimal curing time and under a pressure of 3.94×10^4 kg/m^2. For comparison, the reference flame retardants are a combination of APP and EN (AE) and a combination of APP and GE (AG); the reference composites are neat EPDM, EPDM/AE and EPDM/AG, which were prepared using the same procedure as EPDM/AEG.

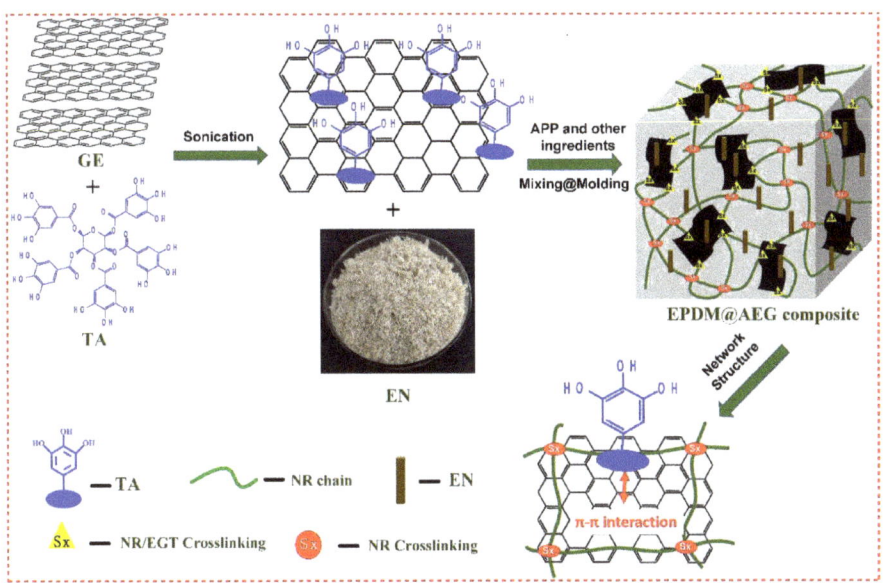

Scheme 1. The preparation process for the flame-retardant EPDM/AEG composites.

Table 1. Batch compositions.

Samples	EPDM (phr)	CB (phr)	ZnO (phr)	SA (phr)	S (phr)	M (phr)	TT (phr)	APP (phr)	EN (phr)	GE (phr)	TA (phr)
Neat EPDM	100	40	5	1	1.5	0.5	1.5	–	–	–	–
EPDM/AEG20	100	40	5	1	1.5	0.5	1.5	20	20	1	0.1
EPDM/AEG30	100	40	5	1	1.5	0.5	1.5	30	20	1	0.1
EPDM/AG40	100	40	5	1	1.5	0.5	1.5	40	–	1	0.1
EPDM/AE40	100	40	5	1	1.5	0.5	1.5	40	20	–	–
EPDM/AEG40	100	40	5	1	1.5	0.5	1.5	40	20	1	0.1

3. Results and Discussion

3.1. The Particle Size Distribution and Zeta Potential of EN

Ultra-fineness is a common flame-retardant modification method that reduces the particle size of the flame retardant to improve the dispersity in the polymer matrix through physical grinding or chemical modification. Through comparing the influence of the mechanical grinding pretreatment method on the distribution of the EN particle size, it can be seen from Figure 1a that the average particle size of EN after the manual grinding treatment is 5218 nm, and that after the mechanical grinding treatment is 2245 nm. Compared with manually ground EN, mechanically ground EN has a narrower particle size distribution and a higher fineness, which gives the EN fibers a better dispersity in the rubber matrix. Moreover, the Zeta potential mainly characterizes the interaction between colloids and electrolytes. The higher the absolute value of the Zeta potential, the more stable the dispersion of the system. The lower the Zeta potential value, the more easily the filler aggregates and condenses. From Figure 1b, it can be seen that the mechanically ground EN has a higher Zeta potential, indicating that it significantly improves the bonding infiltration interaction at the filler–matrix interface.

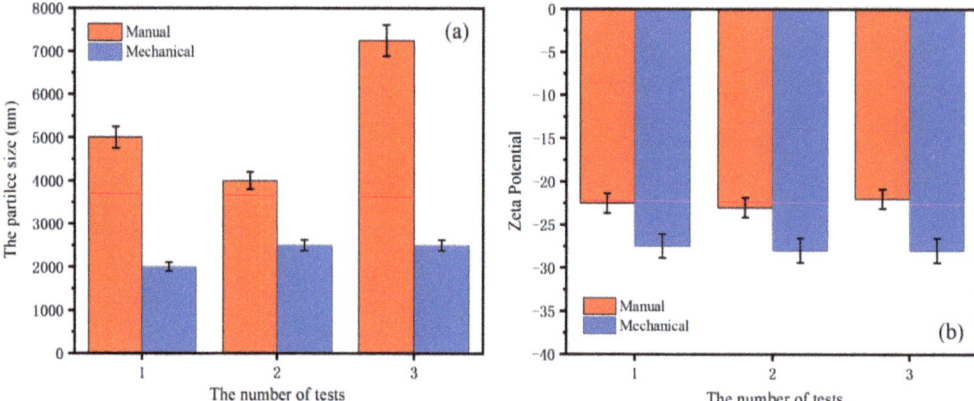

Figure 1. The particle size distribution (**a**) and Zeta potential diagram (**b**) of EN after ultrafine treatment.

3.2. Thermal Stability of EN and AEG40

TGA is used to evaluate the thermal stability of materials, as shown in Figure 2, Figure S1 and Table 2. EN has two obvious peaks at 92.3 °C and 242.5 °C. The appearance of T_{1max} is the decomposition of water and small molecules in the EN biomass. T_{2max} corresponds to the pyrolysis of oxygen-containing functional groups in EN [1,6]. The residual carbon content of EN is relatively high, which can reach 38.1%. Due to the addition of APP, the $T_{-5\%}$ of AEG40 has increased significantly. Because APP decomposes and releases incombustible gas, water and ammonia, the T_{1max} and T_{3max} of AEG40 correspond to the decomposition of polyphosphoric acid to metaphosphoric acid and pyrophosphoric acid to finally generate phosphorous ash. The T_{1max} of AEG40 is higher than the processing temperature of general rubber (150 °C~160 °C), indicating that the flame retardant can maintain good thermal stability during rubber processing without decomposition, and when the rubber burns at 350 °C~450 °C, the flame retardant can quickly decompose to play a flame-retardant effect.

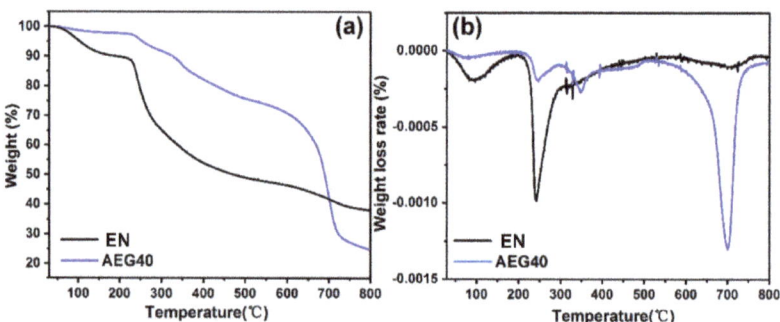

Figure 2. TG curve (**a**) and DTG curve (**b**) of EN and AEG40 under N_2 atmosphere.

Table 2. TG and DTG data of EN and AEG40 under N2 atmosphere.

	$T_{-5\%}$ (°C)	T_{max}(°C)			Residue at 800 °C
		T_{1max} (°C)	T_{2max} (°C)	T_{3max} (°C)	
EN	103.8	92.3	242.5	710.6	38.1
AEG40	261.5	244	249.1	701.7	31.7

Hongsheng Liu et al. [23] have studied that Enteromorpha is not suitable as a filler for conventional polymers, especially for rubber composites, since it is not only is hydrophilic, but also contains some moisture. This moisture is not easy to remove completely, which is due to Enteromorpha growing in salt water. The presence of water is fatal to the vulcanization of rubber, and the vulcanization results of EPDM composites (Table 3) show that Enteromorpha has no effect on the vulcanization and scorching time of EPDM because the important parameters t_{c10} and t_{c90} have not fundamentally changed. The results show that Enteromorpha is almost completely dried.

Table 3. Vulcanization performance EPDM composites.

Samples	ML/dNm^{-1}	MH/dNm^{-1}	MH-ML/dNm^{-1}	tc10/s	t90/s
Neat EPDM	3.10	29.45	26.35	66.6	330.6
EPDM/AEG20	3.38	34.37	30.99	64.2	322.2
EPDM/AEG30	3.61	37.83	34.22	65.4	328.2
EPDM/AG40	3.34	38.29	34.95	64.8	311.4
EPDM/AE40	3.69	37.70	34.01	66.6	312.6
EPDM/AEG40	3.96	38.34	34.38	65.4	313.2

3.3. Flame-Retardant Properties of EPDM Rubber Composites

Neat EPDM burns violently after being ignited, accompanied by severe melting drips. As shown in Table 4, the LOI data show that neat EPDM is flammable because its LOI value is 24.1%. The digital photo of the carbon residue of neat EPDM (Figure 3) directly shows the sparse carbon residue with less carbonization. With the further addition of flame retardants, the LOI value and UL-94 rating of composite materials are elevated. When 40 phr AEG is added, the LOI of EPDM/AEG40 reaches 32.5% and the UL-94 reaches V-0 level. It is worth noting that EPDM/AG40 has the phenomenon of melt dripping. The LOI is 29.5% and the UL-94 level is V-2. EPDM/AE40 has no melt dripping due to the addition of EN, its LOI reaches 31.2% and its UL-94 level reaches V-1. This indicates that EN is of great help in solving melting dripping. We hypothesize that the high carbon-forming property of EN contributes to the formation of a tight protective carbon layer. Nevertheless, the synergistic effect of EN and GE is a good solution to upgrade the flame retardancy grade of EPDM.

To test our conjecture, the cone calorimeter test and a digital photo of the carbon residue of EPDM composites are further studied, as shown in Figure 3. Figure 3 shows that the char residue becomes much denser with the increase in the added AEG. Compared with EPDM/AG40, the amount of carbonization on EPDM/AE40 is much higher, and the char residue is much denser and more complete. This is more pronounced when EN and GE effectively hybridize, as the AEG system can effectively promote the formation of strong carbonization through increasing the degree of aromatic crosslinking. As the main char-forming agent, EN plays an important role in improving the flame-retardant performance. These results can also be verified by thermal analysis.

Table 4. LOI and UL-94 experimental data for EPDM rubber composites.

Samples	t_1	t_2	LOI (%)	UL-94	Dripping
Neat EPDM	45	78	24.1 ± 0.2	NC	Yes
EPDM/AEG20	39	69	28.8 ± 0.4	NC	Yes
EPDM/AEG30	8	12	30.1 ± 0.4	V-1	No
EPDM/AG40	15	20	29.5 ± 0.3	V-2	Yes
EPDM/AE40	13	18	31.2 ± 0.5	V-1	No
EPDM/AEG40	4	10	32.5 ± 0.5	V-0	No

Figure 3. The digital photos of char residues for neat EPDM, EPDM/AEG20, EPDM/AEG30, EPDM/AG40, EPDM/AE40 and EPDM/AEG40 composites after CCT test.

To further study the combustion properties of the AEG system, a more scientific cone calorimeter test was used to evaluate the combustion performance. The burning time after combustion (TTI), the peak heat release rate (PHRR) and total heat release rate (THR), the smoke generation rate (SPR), the total smoke rate (TSR), the total smoke generation (TSP) and the mass residue parameters are shown in Figure 4; the corresponding data are shown in Table 5. Heat is an indispensable stimulant and maintenance factor in a combustion process. For EPDM/AEG composites, AEG flame-retardant systems improve the flame-retardant performances of EPDM mainly by reducing the amount and rate of heat released from combustion. As shown in Figure 4a,b, the peak HRR and THR values have been proven to be key parameters for characterizing the fire safety of polymers. With the increase in the amount of AEG, EPDM/AEG composites show a continuous decrease in HRR and THR. The PHRR and THR values of EPDM/AEG40 are 325.9 kW/m^2 and 117.6 MJ/m^2, respectively, showing a decrement of 67.3% and 29.7%, respectively, compared with the results of the neat EPDM composite. The AEG systems also show superior reduction in combustion heat compared to traditional APP flame retardants. For example, the PHRR and THR values of EPDM/AEG40 reduce by 14.3% and 2.4%, respectively, compared with EPDM/APP40 with 40phr APP. In addition, based on 40 phr APP, AE and GE differ greatly in the degree to which they reduce the amount of generated heat as well as the heat generation rate. For EPDM/AE40, the amount of heat generated and the heat generation rate are 300.75 kW/m^2 of PHRR and 122.1 MJ/m^2 of THR, which are 21.6% and 15.7% lower than EPDM/AG40 (383.4 kW/m^2 of PHRR and 144.8 MJ/m^2 of THR, respectively). These results show that EG contributes significantly to the flame retardancy of GN hybrid materials, especially in reducing combustion heat, which is ascribed to the formation of strong, dense and protective char residue layers.

Table 5. Cone calorimetry experimental data for various samples.

Samples	TTI (s)	PHRR (kW/m^2)	THR (MJ/m^2)	TSP (m^2)	MASS (g)	FIGRA (kW/(m^2/s))	FPI (10^{-2}m^{2}s/kW))	EHC (MJ/kg)
EPDM	34	997.1	167.4	15.2	9.3	3.63	3.41	49.04
EPDM/AEG20	70	460.2	153.5	12.8	16.2	2.01	15.21	40.53
EPDM/AEG30	80	428.1	129.6	10.4	25.4	2.38	18.69	38.84
EPDM/AG40	77	383.4	144.8	11.9	24.3	3.19	20.08	39.07
EPDM/AE40	85	300.75	122.1	10.8	27.6	1.77	28.26	28.62
EPDM/AEG40	97	325.9	117.6	9.9	28.1	1.31	29.76	21.86

Usually, the fire performance index (FPI) [24] and the fire growth index (FIGRA) [25] are used for judging the fire hazard of the studied AEG system more clearly. FPI is defined as the proportion of TTI to PHRR, and there is a certain correlation between the FPI value of a material and its time to flashover. A reduced FPI value means a shorter time to flashover. FIGRA is defined as the proportion of PHRR to the peak HRR time. A larger FIGRA value means a shorter time to reach the peak HRR time and the greater the fire hazard is for the materials. The FPI and FIGRA values of the studied composites have been shown in Table 5. Based on 40 phr of the main flame retardant APP, the synergistic flame retardancy and smoke suppression effects of GN, EN and GE on EPDM are compared. As can be seen from Table 5, when EN and GE are effectively hybridized, the fire risk of GN is small, which is equivalent to a low fire risk. This is because GN combustion produces a large amount of residual char with a high degree of graphitization, thus forming a dense carbon layer. Moreover, the residual char produced by EN is fibrous and effectively cross-linked with the aromatic residue char contributed by graphene to strengthen the carbon layer.

Usually in a fire, most fire casualties are caused by asphyxiation caused by smoke, so it is very important to reduce smoke generation and emission during combustion [25]. The dynamic smoke generation behavior of EPDM composites is characterized by total smoke generation (TSP), the total smoke rate (TSR) and the smoke generation rate (RSR), as shown in Figure 4c–e. The data are summarized in Table 5. As shown in Figure 4c, compared with EPDM, the TSP, TSR and RSR values of EPDM/AEG composites are significantly reduced. The TSP and TSR values of EPDM/AEG40 are especially reduced from 15.2 m^2 for neat EPDM to 9.9 m^2 for TSP, with a decrement of 34.9%, and reduced from 1715.2 m^2/m^2 for neat EPDM to 1124.5 m^2/m^2 for TSR, with a decrement of 34.4%, indicating that AEG is an effective flame retardant and smoke suppressor. Evidently, EN has an obvious effect on reducing the smoke release rate, resulting in a reduced TSP and TSR. The P-SPR of EPDM/AEG40 composite is also 69.4% lower than that of neat EPDM. It can be seen that the AEG flame-retardant system has an obvious smoke suppression effect by acting as a physical barrier to limit the transfer of smoke and dust. The slow release of heat and smoke is very beneficial for fire control and the escape of people caught in the fire. In addition, compared with neat EPDM, the reductions in the carbon monoxide and carbon dioxide generation rates of EPDM/AEG composites (Figure 5) also directly prove the higher char-forming ability of AEG. Notably, the peak release of CO for the EPDM/AEG40 is just 0.0057 g/s. This is a considerable flame-retardant index because CO can cause suffocation in a fire, which is the most common cause of death in a fire. As a result of high emissions of carbon dioxide, global warming is presenting a range of harms to humans, and AEG shows excellent performance in inhibiting CO_2 release. The measured average effective hot comb (EHC) further provides a better understanding of the reduced flammability caused by coke formation [26]. The thermal combustion performance of EPDM/AEG composites (Table 5) proves the synergistic improvement in thermal stability and flame retardancy. In this regard, we conclude that GN acts as a carbon source in the AEG system and plays a very good role in preventing heat and flammable volatiles in the condensed phase.

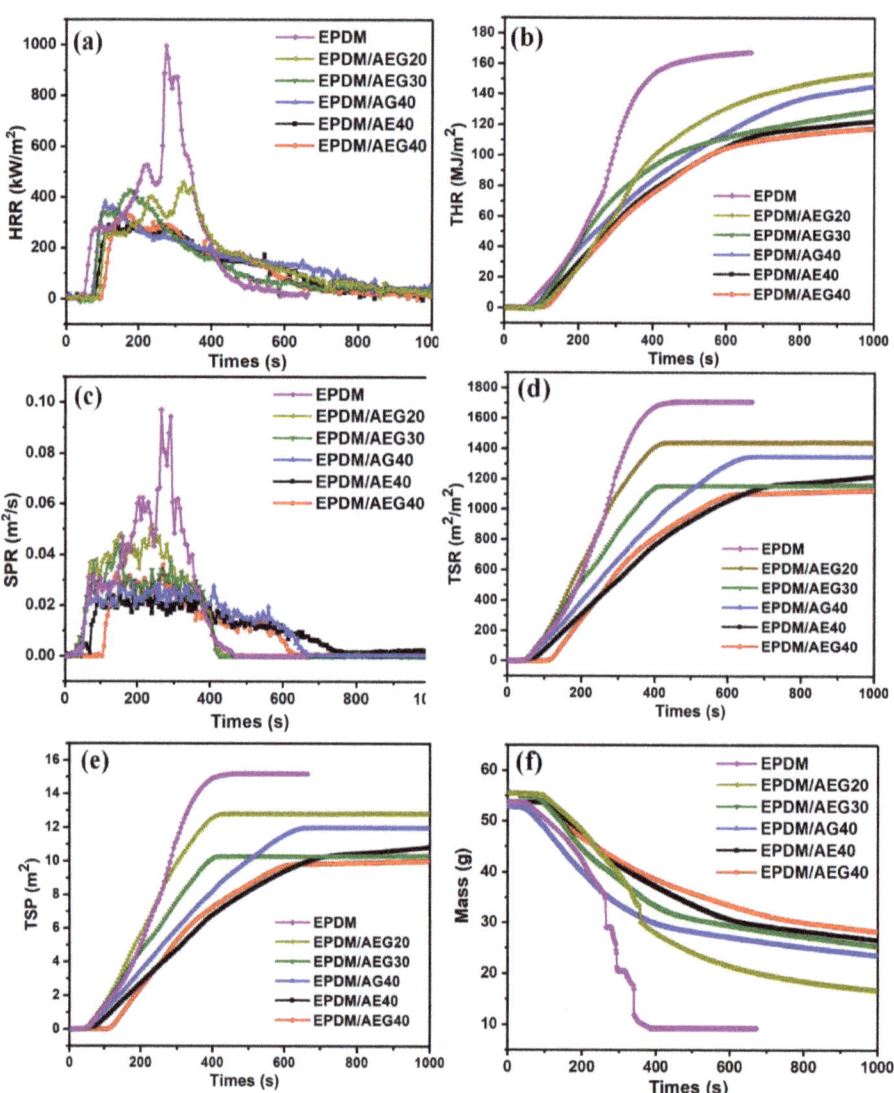

Figure 4. CCT results of EPDM composites with 35 kW/m² external heat flux: (**a**) HRR, (**b**) THR, (**c**) RSR, (**d**) TSR, (**e**) TSP and (**f**) MASS.

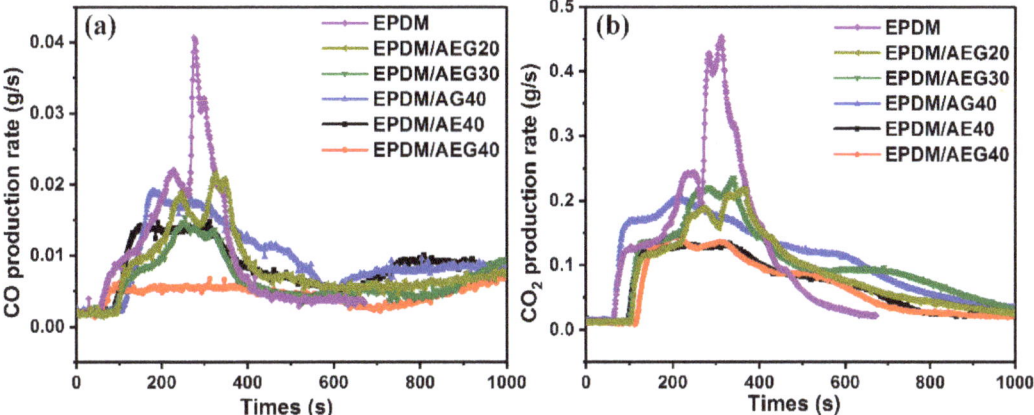

Figure 5. CCT results of EPDM composites under an external heat flux of 35 kW/m^2: (**a**) released CO production rate and (**b**) released CO$_2$ production rate.

3.4. Morphology of Intumescent Char Layer

In order to clarify the influence of char residue on the combustion of EPDM composite materials, the morphology of the residual chars after combustion are studied by SEM. Figure 6 shows the SEM images of the char residues of neat EPDM, EPDM/AG, EPDM/AEG and EPDM/AE composites after the cone calorimeter test. For the char residue of neat EPDM, a loose char layer can be clearly observed, along with no expansion, slight melting and shedding (Figure 6a). This is due to the lack of a char-forming agent, the insufficient thickness and the unbreakable strength of the carbon residue. The char residue of EPDM/AEG composites (Figure 6b,c,f) is more compact and complete, and with the increase in the amount of APP, the increase in acid source makes the char layer expand more obviously. Furthermore, to learn more about flame-retardant effect of EN and GE combined use, the char residues of EPDM/AG40, EPDM/AE40 and EPDM/AEG40 are studied by SEM images (Figure 6d–f). For the residues of EPDM/AG40 and EPDM/AE40, there is a large number of cracks and visible holes, which have a unavoidable negative effect on the barrier action of these residues; however, the residue of EPDM/AEG40 is more continuous, and there is almost no visible holes. Under high magnification conditions (Figure 6(d-2,e-2,f-2)), the degree of contrast is more obvious. For EPDM/AEG40, not only are the flammable and oxygen gasses blocked from the flame by its superior char layer, but the heat transfer process is prevented, resulting in a better flame retardancy for the EPDM/AEG40.

SEM-EDS analysis was used to study the distribution and content of flame-retardant elements in the carbon residue. In Figure 7, the element mapping of the EPDM/AEG composites (Figure 7b,d,f) reveals changes in the content of carbon, nitrogen, oxygen and phosphorus atoms on the carbon surface. As shown in Table 6, compared with neat EPDM, EPDM/AEG composites have an increased carbon atom content and a decreased oxygen atom content, which further illustrates the ability of AEG to promote residue formation during combustion. The increase in phosphorus atoms further confirms the high phosphorus residue after combustion, which provides higher resistance during pyrolysis and combustion.

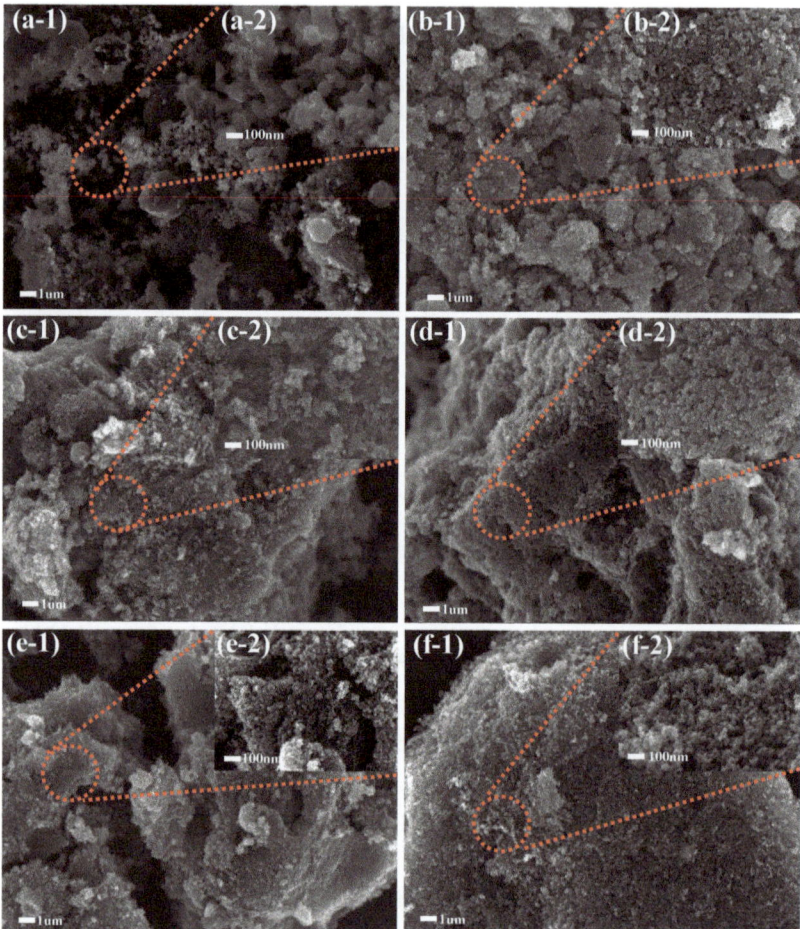

Figure 6. Microstructure of the residue char after CCT test at different resolution: (**a1,a2**) neat EPDM, (**b1,b2**) EPDM/AEG20, (**c1,c2**) EPDM/AEG30, (**d1,d2**) EPDM/AG40, (**e1,e2**) EPDM/AE40, (**f1,f2**) EPDM/AEG40.

Table 6. Element content of EPDM in the SEM-EDS spectrum.

Samples	C (Weight, %)	N (Weight, %)	O (Weight, %)	P (Weight, %)
Neat EPDM	57.91	0.93	31.64	2.38
EPDM/AEG20	72.78	2.44	17.70	4.18
EPDM/AEG30	74.13	2.16	19.63	4.08
EPDM/AG40	77.39	2.24	15.85	4.05
EPDM/AE40	79.38	2.61	14.06	3.96
EPDM/AEG40	86.22	2.75	6.31	4.72

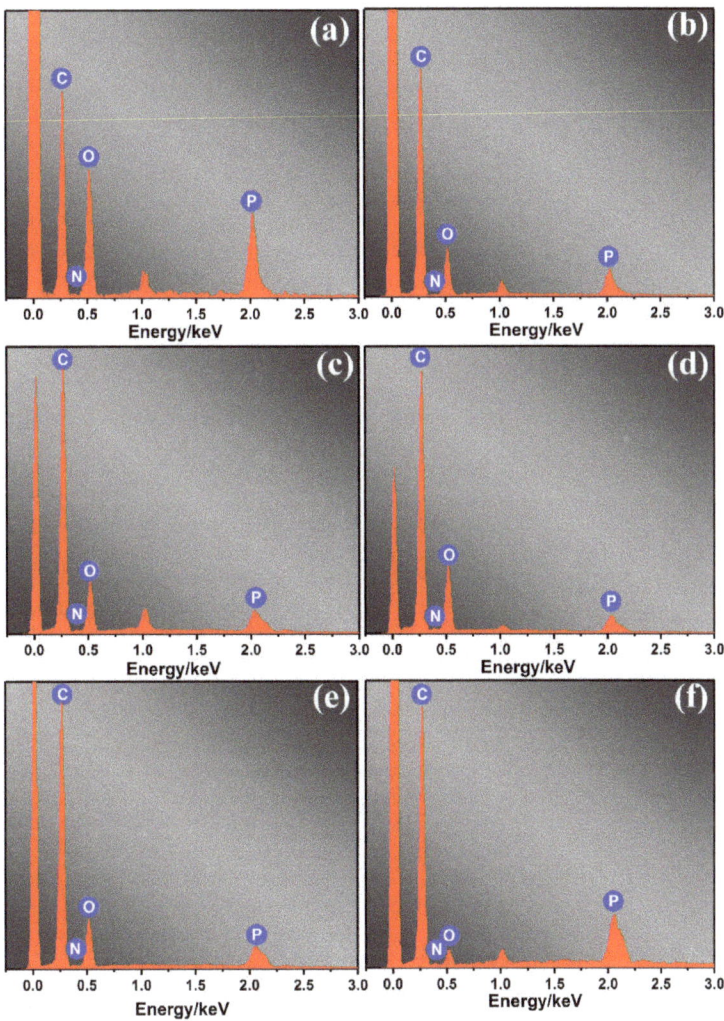

Figure 7. SEM-EDS spectra of (**a**) neat EPDM, (**b**) EPDM/AEG20, (**c**) EPDM/AEG30, (**d**) EPDM/AG40, (**e**) EPDM/AE40, (**f**) EPDM/AEG40.

3.5. Static and Dynamic Mechanical Properties of EPDM/AEG Composites

The physical properties are a non-ignorable performance index for elastic rubber in almost all applications [27]. It should be emphasized that, unfortunately, there is a tendency for most flame retardants to remarkably damage the mechanical behavior of rubber, especially for nonpolar rubber. In most cases, polarity differences between polar fire retardants and nonpolar rubber result in the poor particle dispersion and compatibility that are responsible for these negative effects. However, an abnormal phenomenon occurs in the AEG system in that AEG has a positive effect on the rubber's mechanical properties, which is an urgent problem of the commonly used flame retardants needing to be solved. As shown in Figure 8, tensile strength and tear strength of EPDM/AEG composites are much higher than neat EPDM and EPDM/APP40, even for EPDM/AEG40 with a V-0 flame retardant rating, which has a tensile strength up to 33.0% and a tear strength up to 30.1% based on neat EPDM. However, while the tensile strength and tear strength of the EPDM/AEG composites both show a tendency to decrease increase in the APP content, they are always higher than the neat EPDM. For, EPDM/APP40 with a V-0 flame retardant rating, the tensile strength goes down to 29.8% compared to neat EPDM (Table S4). Obviously, GN can compensate for the loss of mechanical properties caused by APP. Under the load of 40 phr APP, the tensile strength for EPDM/AE40 and EPDM/AG40 is also improved, resulting in EPDM/AE40 and EPDM/AG40 being 23.5% and 12.1% higher than neat EPDM, respectively. Moreover, what is more exciting is that AEG has a toughening effect as well as a strengthening effect on EPDM. By contrast, after effective hybridization between EN and GE, the tensile strength and tear strength of EPDM/AEG40 are both superior to EPDM/AE40 and EPDM/AG40, which is attributed to the synergistic effect of AEG caused by the reinforcing effect of the two-dimensional graphene with a high specific surface area and the one-dimensional Entermorpha fiber with a high aspect ratio. In order to explore the dynamic mechanical properties of EPDM/AEG composites, dynamic mechanical analysis was used. Figure 9a shows that the addition of AGE has little effect on the glass transition temperature (T_g) of vulcanized EPDM. It is well dispersed and has little effect on the movement and internal friction of the EPDM molecular chain. According to Figure 9b, the initial values of E' of EPDM/AEG30 and EPDM/AEG40 are significantly higher than those of EPDM/AG40 and EPDM/AE40, indicating that AEG has stronger reinforcing ability than AG and AE systems. It can also be seen that the storage modulus of EPDM increases with the increase in filler, which is attributed to the reinforcing effect of EN. The tan delta values at 0 °C of EPDM/AEG30 and EPDM/AEG40 are similar to EPDM/AE40 but significantly higher than that of EPDM/AG40, which means EPDM/AEG30, EPDM/AEG40 and EPDM/AE40 have better wet-skid resistance. The results indicate that the contribution of Enteromorpha in flame-retardant systems is critical to the wet-skid resistance of NR composites [28].

3.6. Flame Retardancy Mechanism

Based on the above analysis, the flame retardancy mechanism of the AEG system in the EPDM matrix is assumed in Figure 10. First of all, for the condensed phase, in the early degradation process, the premature decomposition of EN absorbs part of the heat, which causes the ignition time of the polymer to be delayed [28]. GE sheets with excellent thermal stability act as quality barrier to inhibit the penetration of combustible gas [29,30]. At the same time, the catalytic carbonization of metaphosphoric acid (HPO_4) decomposed by the APP adsorbed on the graphene surface leads to the production of a great amount of phosphorous and carbonaceous carbon residues [31,32], which is also confirmed from the EDS data shown in Figure 7 and Table 5. In addition, since EN maintains a good fiber morphology after burning, the high aspect ratio of EN can hold the carbon particles together, and the ammonia gas released from the APP expands to form a honeycomb carbon layer to produce high strength on the internal material. Because of its relatively large specific surface area, TGE can play a role in reinforcing the carbon layer and capturing free radicals in the gas phase. Secondly, with regard to the gas phase, the ammonia decomposed from APP effectively dilutes the volatile during combustion. It can be seen that the tripartite cooperation mechanism of the AEG system is the main source of the excellent flame-retardant properties of EPDM composites. In order to confirm this hypothesis, the microstructure of coke slag was analyzed by Raman spectroscopy.

Raman spectroscopy reveals the flame retardancy mechanism of AEG. Figure 11 and Table S3 show that peak fitting is performed for each spectrum to resolve the curve into D and G bands (Figure 11). The combined intensity (I_D/I_G) reflects the graphitization degree of residual carbon, and the lower the I_D/I_G value is, the better the residual carbon structure is [33–35]. Obviously, the I_D/I_G ratio in this study follows the sequence of EPDM/AEG40 < EPDM/AE40 < EPDM/AG40, which indicates that EN can promote the formation of highly graphitized and insulating carbon layers, which is the main mechanism for suppressing flammability.

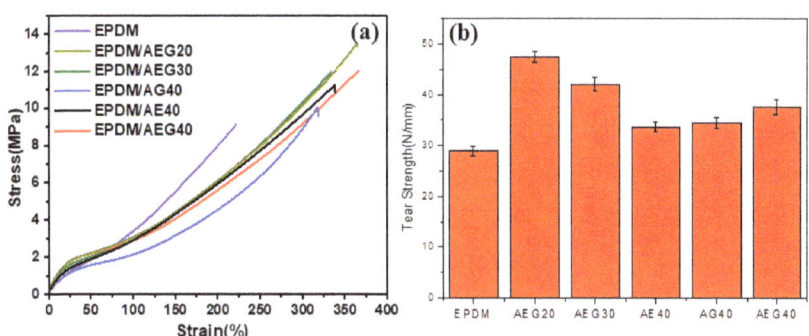

Figure 8. (**a**) Tensile strength and (**b**) tear strength of EPDM/AEG.

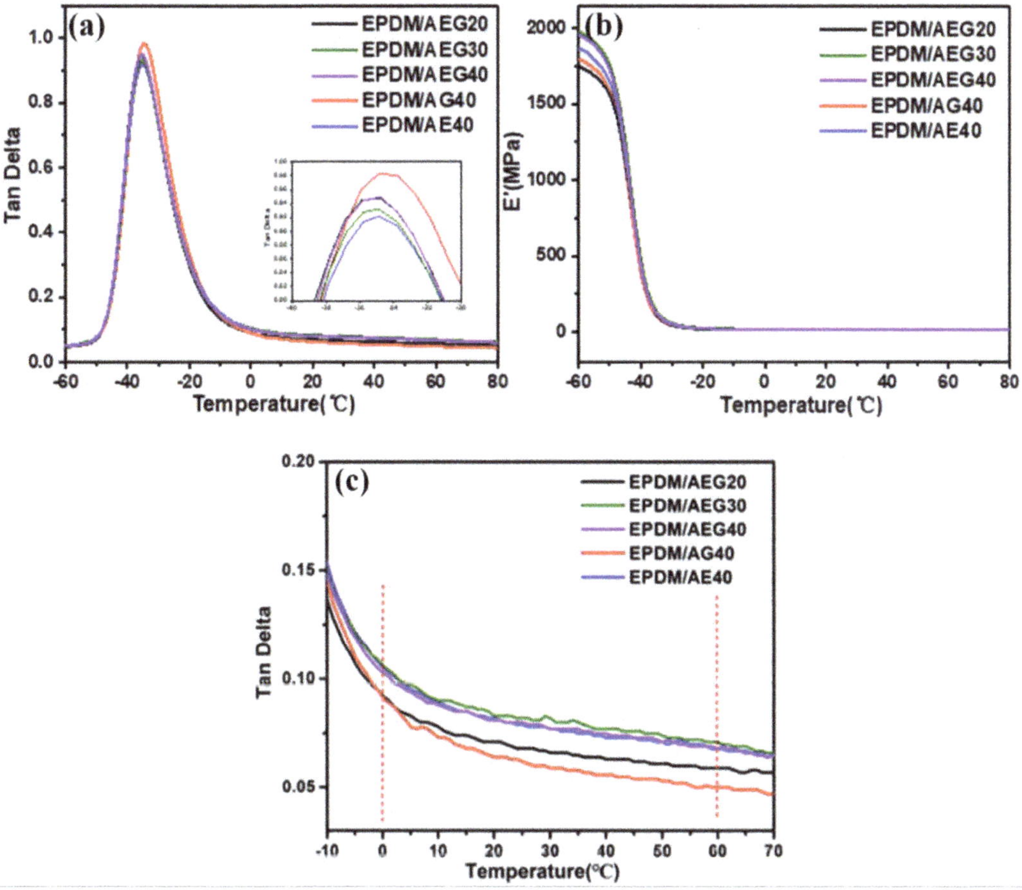

Figure 9. Tan δ curve (**a**,**c**) and storage modulus curve (**b**) of EPDM/AEG composites.

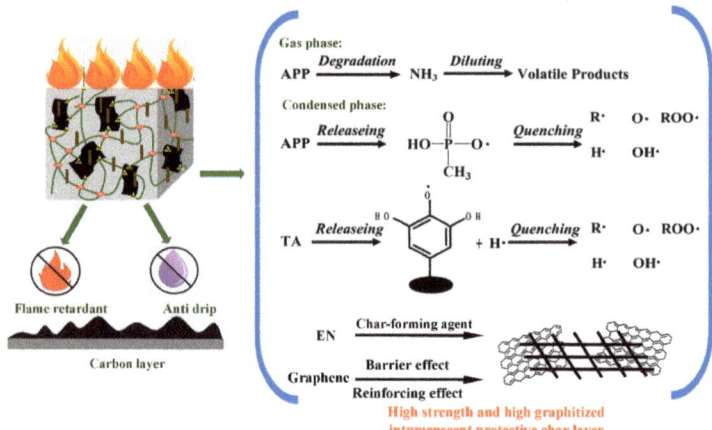

Figure 10. Illustrative scheme of the flame retardancy mechanism for EPDM/AEG composites.

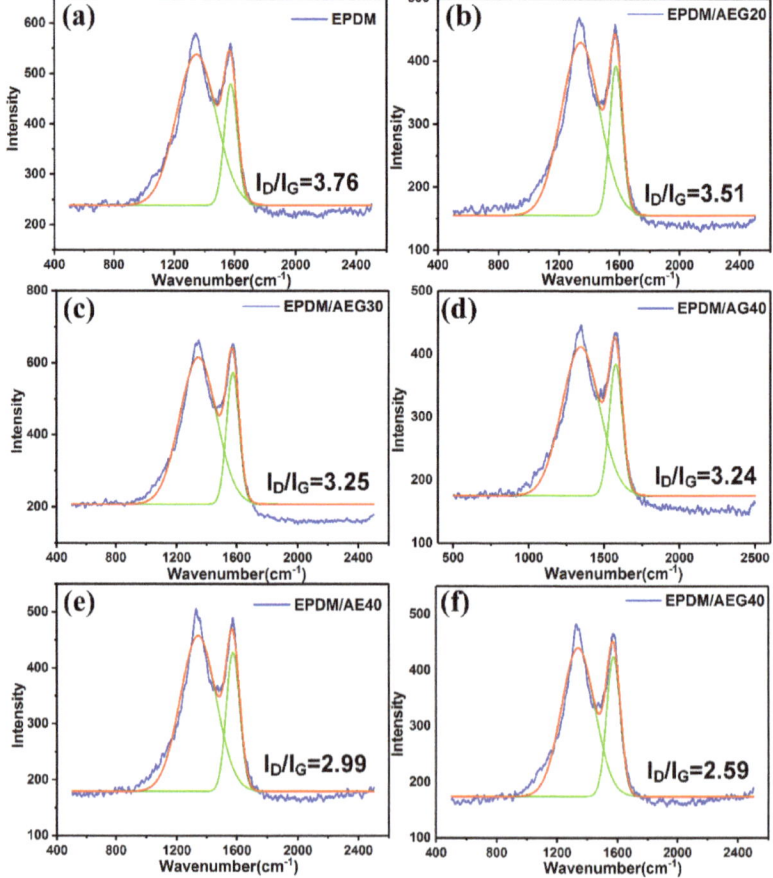

Figure 11. Raman curves of residue chars of (**a**) EPDM, (**b**) EPDM/AEG20, (**c**) EPDM/AEG30, (**d**) EPDM/AG40, (**e**) EPDM/AE40 and (**f**) EPDM/AEG40 composites.

4. Conclusions

The bio-based AEG flame retardants has a synergistic effect on smoke suppression and flame retardancy in EPDM due to the intumescent flame retardancy mechanism. The microstructure obtained from Raman and SEM-EDS analyses of the residual chars show that the AEG can significantly promote high graphitization and contains a phosphorus structure. Thus, flammability parameters and fire risk are effectively reduced. In addition, the AEG is beneficial for the mechanical properties of EPDM within a proper addition level. Even under high load conditions, EN and GE can compensate for the negative effects caused by the addition of flame retardants on mechanical properties.

Supplementary Materials: The following supporting information can be downloaded at: https://www.mdpi.com/article/10.3390/polym15010055/s1, Figure S1: TG curve (a) and DTG curve (b) of APP under N_2 atmosphere; Table S1: The batch compositions for AEG system title; Table S2: LOI tests experimental data for various samples; Table S3: Cone calorimetry experimental data for EPDM/APP40; Table S4: Mechanical properties for EPDM/APP40; Table S5: Raman results of EPDM/AEG cofimposites.

Author Contributions: Conceptualization, Z.Z.; data curation, L.J.; writing—original draft preparation, P.S.; writing—review and editing, L.L.; supervision, L.L.; project administration, S.Z.; funding acquisition, P.S., L.L. and S.Z. All authors have read and agreed to the published version of the manuscript.

Funding: Financial support of this project was funded by the Natural Science Foundation of China (51703111 and 51603111), the Innovation Pilot Project of Integration of Science, Education and Industry of Shandong Province (2020KJC-ZD06), the Natural Science Foundation of Shandong Province (ZR2021ME107) and the project funded by the China Postdoctoral Science Foundation (2021M700553 and 2020M672014) are gratefully acknowledged. The project was also supported by the Foundation (ZR20200101) of the State Key Laboratory of Biobased Material and Green Papermaking, Qilu University of Technology, Shandong Academy of Sciences.

Informed Consent Statement: The studies do not involve humans.

Acknowledgments: Financial support is provided by the Natural Science Foundation of China (51703111 and 51603111), the Natural Science Foundation of Shandong Province (ZR2021ME107) and the China Postdoctoral Science Foundation (2022M721903, 2021M700553 and 2020M672014). The project is also supported by the Foundation (ZR20200101) of the State Key Laboratory of Biobased Material and Green Papermaking, Qilu University of Technology, Shandong Academy of Sciences.

Conflicts of Interest: The authors declare no conflict of interest.

References

1. Ji, K.; Gao, Y.; Zhang, L.; Wang, S.; Yue, Q.; Xu, X.; Kong, W.; Gao, B.; Cai, Z.; Chen, Y. A tunable amphiphilic Enteromorpha-modified graphene aerogel for oil/water separation. *Sci. Total Environ.* **2021**, *763*, 142958. [CrossRef] [PubMed]
2. Chen, C.; Ma, T.; Shang, Y.; Gao, B.; Jin, B.; Dan, H.; Li, Q.; Yue, Q.; Li, Y.; Wang, Y.; et al. In situ pyrolysis of Enteromorpha as carbocatalyst for catalytic removal of organic contaminants: Considering the intrinsic N/Fe in Enteromorpha and non-radical reaction. *Appl. Catal. B Environ.* **2019**, *250*, 382–395. [CrossRef]
3. Zhong, R.T.; Wan, X.Z.; Wang, D.Y.; Zhao, C.; Liu, D.; Gao, L.Y.; Wang, M.F.; Wu, C.J.; Sayed, M.N.; Maria, D.L.; et al. Polysaccharides from marine enteromorpha: Structure and function. *Trends Food Sci. Technol.* **2020**, *99*, 11–20. [CrossRef]
4. Li, Y.; Cui, J.; Zhang, G.; Liu, Z.; Guan, H.; Hwang, H.; Aker, W.G.; Wang, P. Optimization study on the hydrogen peroxide pretreatment and production of bioethanol from seaweed Ulva prolifera biomass. *Bioresour. Technol.* **2016**, *214*, 144–149. [CrossRef] [PubMed]
5. Wang, Y.; Yu, L.; Xie, F.; Zhang, L.; Liao, L.; Liu, H.; Chen, L. Morphology and properties of thermal/cooling-gel bi-phasic systems based on hydroxypropyl methylcellulose and hydroxypropyl starch. *Compos. Part B Eng.* **2016**, *101*, 46–52. [CrossRef]
6. Duan, Q.; Jiang, T.; Xue, C.; Liu, H.; Liu, F.; Alee, M.; Ali, A.; Chen, L.; Yu, L. Preparation and characterization of starch/enteromorpha/nano-clay hybrid composites. *Int. J. Biol. Macromol.* **2020**, *150*, 16–22. [CrossRef]
7. Pradhan, B.; Patra, S.; Behera, C.; Nayak, R.; Jena, M. Preliminary investigation of the antioxidant, anti-diabetic, and anti-inflammatory activity of enteromorpha intestinalis extracts. *Molecules* **2021**, *26*, 1171. [CrossRef]
8. Shanab, S.M.M.; Shalaby, E.A. The role of salt stress on laboratory cultivation of green macroalga Enteromorpha compressa and its antioxidant activity. *Baghdad Sci. J.* **2020**, *18*, 54. [CrossRef]

9. Wassie, T.; Niu, K.; Xie, C.; Wang, H.; Xin, W. Extraction Techniques, Biological Activities and Health Benefits of Marine Algae Enteromorpha prolifera Polysaccharide. *Front. Nutr.* **2021**, *8*, 747928. [CrossRef]
10. Li, J.; Jiang, F.; Chi, Z.; Han, D.; Yu, L.; Liu, C. Development of Enteromorpha prolifera polysaccharide-based nanoparticles for delivery of curcumin to cancer cells. *Int. J. Biol. Macromol.* **2018**, *112*, 413–421. [CrossRef]
11. Wassie, T.; Duan, X.; Xie, C.; Wang, R.; Wu, X. Dietary Enteromorpha polysaccharide-Zn supplementation regulates amino acid and fatty acid metabolism by improving the antioxidant activity in chicken. *J. Anim. Sci. Biotechnol.* **2022**, *13*, 1–19. [CrossRef] [PubMed]
12. Ray, B.; Schütz, M.; Mukherjee, S.; Jana, S.; Ray, S.; Marschall, M. Exploiting the Amazing Diversity of Natural Source-Derived Polysaccharides: Modern Procedures of Isolation, Engineering, and Optimization of Antiviral Activities. *Polymers* **2020**, *13*, 136. [CrossRef] [PubMed]
13. Venkatesan, M.; Arumugam, V.; Pugalendi, R.; Ramachandran, K.; Sengodan, K.; Vijayan, S.R.; Sundaresan, U.; Ramachandran, S.; Pugazhendhi, A. Antioxidant, anticoagulant and mosquitocidal properties of water soluble polysaccharides (WSPs) from Indian seaweeds. *Process. Biochem.* **2019**, *84*, 196–204. [CrossRef]
14. Ouyang, Y.; Liu, D.; Zhang, L.; Li, X.; Chen, X.; Zhao, C. Green Alga *Enteromorpha prolifera* Oligosaccharide Ameliorates Ageing and Hyperglycemia through Gut–Brain Axis in Age-Matched Diabetic Mice. *Mol. Nutr. Food Res.* **2022**, *66*, 2100564. [CrossRef] [PubMed]
15. Cui, J.; Xi, Y.; Chen, S.; Li, D.; She, X.; Sun, J.; Han, W.; Yang, D.; Guo, S. Prolifera-Green-Tide as Sustainable Source for Carbonaceous Aerogels with Hierarchical Pore to Achieve Multiple Energy Storage. *Adv. Funct. Mater.* **2016**, *26*, 8487–8495. [CrossRef]
16. Pitchapa, P.; Sombat, T.; Taweechai, A. Comparative study of natural rubber and acrylonitrile rubber reinforced with aligned short aramid fiber. *Polym. Test.* **2017**, *64*, 109–116.
17. Lee, M.C.H. *Design and Applications of Short Fiber Reinforced Rubbers*; De White, S.K., Ed.; Short Fiber-Polymer Composites, Woodhead Publishing: Cambridge, UK, 1996; pp. 192–209.
18. Agarwal, K.; Setua, D.K.; Mathur, G.N. Short fiber and particulate reinforced rubber composites. *Def. Sci. J.* **2002**, *52*, 337–346. [CrossRef]
19. Yang, G.; Qin, L.; Li, M.; Ou, K.; Fang, J.; Fu, Q.; Sun, Y. Shear-induced alignment in 3D-printed nitrile rubber-reinforced glass fiber composites. *Compos. Part B Eng.* **2022**, *229*, 109479. [CrossRef]
20. Tang, L.; He, M.; Na, X.; Guan, X.; Zhang, R.; Zhang, J.; Gu, J. Functionalized glass fibers cloth/spherical BN fillers/epoxy laminated composites with excellent thermal conductivities and electrical insulation properties. *Compos. Commun.* **2019**, *16*, 5–10. [CrossRef]
21. Lu, S.-Y.; Hamerton, I. Recent developments in the chemistry of halogen-free flame retardant polymers. *Prog. Polym. Sci.* **2002**, *27*, 1661–1712. [CrossRef]
22. Bulota, M.; Budtova, T. PLA/algae composites: Morphology and mechanical properties. *Compos. Part A Appl. Sci. Manuf.* **2015**, *73*, 109–115. [CrossRef]
23. Shao, Z.-B.; Deng, C.; Tan, Y.; Yu, L.; Chen, M.-J.; Chen, L.; Wang, Y.-Z. Ammonium polyphosphate chemically-modified with ethanolamine as an efficient intumescent flame retardant for polypropylene. *J. Mater. Chem. A* **2014**, *2*, 13955–13965. [CrossRef]
24. Wu, J.-N.; Chen, L.; Fu, T.; Zhao, H.-B.; Guo, D.-M.; Wang, X.-L.; Wang, Y.-Z. New application for aromatic Schiff base: High efficient flame-retardant and anti-dripping action for polyesters. *Chem. Eng. J.* **2018**, *336*, 622–632. [CrossRef]
25. Li, L.; Liu, X.; Shao, X.; Jiang, L.; Huang, K.; Zhao, S. Synergistic effects of a highly effective intumescent flame retardant based on tannic acid functionalized graphene on the flame retardancy and smoke suppression properties of natural rubber. *Compos. Part A Appl. Sci. Manuf.* **2019**, *129*, 105715. [CrossRef]
26. Lu, Y.; Liu, J.; Hou, G.; Ma, J.; Wang, W.; Wei, F.; Zhang, L. From nano to giant? Designing carbon nanotubes for rubber reinforcement and their applications for high performance tires. *Compos. Sci. Technol.* **2016**, *137*, 94–101. [CrossRef]
27. Zheng, X.; Song, S.K.; Zhou, Z.; Jiang, L.; Sui, Y.; Che, M.; Xu, Q.; Wang, Y.; Zhao, S.; Li, L. Effect of silica dispersed by special dispersing agents with green strategy on tire rolling resistance and energy consumption. *J. Appl. Polym. Sci.* **2022**, *139*, e52933. [CrossRef]
28. Jing, J.; Zhang, Y.; Fang, Z.-P.; Wang, D.-Y. Core-shell flame retardant/graphene oxide hybrid: A self-assembly strategy towards reducing fire hazard and improving toughness of polylactic acid. *Compos. Sci. Technol.* **2018**, *165*, 161–167. [CrossRef]
29. Guo, W.; Yu, B.; Yuan, Y.; Song, L.; Hu, Y. In situ preparation of reduced graphene oxide/DOPO-based phosphonamidate hybrids towards high-performance epoxy nanocomposites. *Compos. Part B Eng.* **2017**, *123*, 154–164. [CrossRef]
30. Xu, W.; Wang, X.; Wu, Y.; Li, W.; Chen, C. Functionalized graphene with Co-ZIF adsorbed borate ions as an effective flame retardant and smoke suppression agent for epoxy resin. *J. Hazard. Mater.* **2019**, *363*, 138–151. [CrossRef]
31. Khalili, P.; Liu, X.; Tshai, K.Y.; Rudd, C.; Yi, X.; Kong, I. Development of fire retardancy of natural fiber composite encouraged by a synergy between zinc borate and ammonium polyphosphate. *Compos. Part B Eng.* **2019**, *159*, 165–172. [CrossRef]
32. Dresselhaus, M.; Dresselhaus, G.; Jorio, A.; Souza Filho, A.G.; Saito, R. Raman spectroscopy on isolated single wall carbon nanotubes. *Carbon* **2002**, *40*, 2043–2061. [CrossRef]
33. Tuinstra, F.; Koening, J.L. Raman spectrum of graphite. *J. Chem. Phys.* **1970**, *53*, 1126–1130. [CrossRef]

34. Wang, J.; Wang, X.; Zhou, Z.; Liu, X.; Xu, M.; Zhao, F.; Zhao, F.; Li, S.; Liu, Z.; Li, L.; et al. Flame-retardant effect of tannic acid-based intumescent fire-retardant applied on flammable natural rubber. *RSC Adv.* **2022**, *12*, 29928–29938. [CrossRef] [PubMed]
35. Ferrari, A.C.; Robertson, J. Interpretation of Raman spectra of disordered and amorphous carbon. *Phys. Rev. B* **2000**, *61*, 14095–14107. [CrossRef]

Disclaimer/Publisher's Note: The statements, opinions and data contained in all publications are solely those of the individual author(s) and contributor(s) and not of MDPI and/or the editor(s). MDPI and/or the editor(s) disclaim responsibility for any injury to people or property resulting from any ideas, methods, instructions or products referred to in the content.

MDPI
St. Alban-Anlage 66
4052 Basel
Switzerland
www.mdpi.com

Polymers Editorial Office
E-mail: polymers@mdpi.com
www.mdpi.com/journal/polymers

Disclaimer/Publisher's Note: The statements, opinions and data contained in all publications are solely those of the individual author(s) and contributor(s) and not of MDPI and/or the editor(s). MDPI and/or the editor(s) disclaim responsibility for any injury to people or property resulting from any ideas, methods, instructions or products referred to in the content.

www.ingramcontent.com/pod-product-compliance
Lightning Source LLC
LaVergne TN
LVHW070624100526
838202LV00012B/722